"Whether it's AI, demographic shifts, or dramatic changes to employee expectations, many forces are transforming employment, careers, and jobs. This book provides a compelling illustration of this new order of things, as well as an exciting vision for the future."

—**TOMAS CHAMORRO-PREMUZIC,** Professor of Business Psychology, University College London and Columbia University; author, *I, Human*

"*Employment Is Dead* presents a compelling case for moving beyond traditional employment models to embrace decentralized, flexible work environments. It's a must-read for leaders and professionals eager to harness the power of these innovations and redefine what it means to work in the modern era."

—**JAMES CROYLE,** CEO, MetaHub Finance

"*Employment Is Dead* is a must-read for anyone ready to shake up the old HR playbook and create a workplace where everyone thrives. This is a revolutionary approach to work that taps into the heartbeat of every single individual to create a more human, joyful, and effective organization."

—**CLAUDE SILVER,** Chief Heart Officer, *VaynerX*

EMPLOYMENT IS DEAD

EMPLOYMENT IS DEAD

HOW DISRUPTIVE TECHNOLOGIES ARE REVOLUTIONIZING THE WAY WE WORK

DEBORAH PERRY PISCIONE AND JOSH DREAN

HARVARD BUSINESS REVIEW PRESS
BOSTON, MASSACHUSETTS

Printed in the United States of America

10 9 8 7 6 5 4 3 2 1

Library of Congress Cataloging-in-Publication Data

Names: Piscione, Deborah Perry, author. | Drean, Josh, author.
Title: Employment is dead : how disruptive technologies are revolutionizing the way we work / Deborah Perry Piscione and Josh Drean.
Description: Boston, Massachusetts : Harvard Business Review Press, [2025] | Includes index. |
Identifiers: LCCN 2024021282 (print) | LCCN 2024021283 (ebook) | ISBN 9781647826420 (hardcover) | ISBN 9781647826437 (epub)
Subjects: LCSH: Employees—Effect of technological innovations on. | Employment (Economic theory)
Classification: LCC HD6331 .P567 2025 (print) | LCC HD6331 (ebook) | DDC 658.3/01—dc23/eng/20240819
LC record available at https://lccn.loc.gov/2024021282
LC ebook record available at https://lccn.loc.gov/2024021283

ISBN: 978-1-64782-642-0
eISBN: 978-1-64782-643-7

The paper used in this publication meets the requirements of the American National Standard for Permanence of Paper for Publications and Documents in Libraries and Archives Z39.48-1992.

CONTENTS

PREFACE

As business leaders grapple with the forces shaping the future of work, it becomes clear that we are on the threshold of a transformation that is both inevitable and profound. The conventional work structures that have long governed our professional lives are being upended. Disruptive technologies such as artificial intelligence (AI), decentralized systems, and web3 technologies are extraordinary technological shifts that are rewriting the social contract that has underpinned our organizations for more than a century. The question arises, then, Is traditional employment dead?

The confines of hierarchical management and centralized control are increasingly ill-suited for the dynamic, fast-paced ecosystems emerging around us. What is taking their place? Decentralized work ecosystems that empower individuals with unprecedented flexibility and agency to mix and match income streams, collaborate in communities driven by shared passions, and engage in meaningful work from anywhere in the world. In this new vessel for work, there are unparalleled opportunities for individual ownership, autonomy, and participative governance fueled by the hyper gig economy of project-based work. In this future, work has evolved into a ceaseless, global orchestration of collaborative efforts where workers spend the majority of time on their passions, and far less time on the monotonous tasks assigned to them.

Yet this newfound revolution comes with its own set of challenges for the corporate leader and the future of work—a provocation that calls for adaptability and forward thinking. The tools of tomorrow, from AI algorithms to digital twin mirroring, for example, are the instruments that will amplify human potential. Those who embrace these technologies and the manner in which people want to work will unlock unprecedented levels of productivity and innovation. Conversely, those who remain tethered to outdated work patterns risk obsolescence.

This book argues that traditional employment models are becoming obsolete, giving way to a decentralized, human-centric future of work powered by disruptive technologies like AI, blockchain, and web3. We will explore an uncharted territory teeming with potential to make work opportunities better, more human and democratized. Decentralization will offer workers a whole new landscape that serves to empower rather than constrain.

Are you prepared to engage in the next evolution of work?

INTRODUCTION

Employment Is Dead

Artificial intelligence. The metaverse. Blockchain. Virtual reality. The conversation around how disruptive technologies will impact work is often polarized. While some leaders are passionate advocates for technology's transformative power, others view the hype surrounding these innovations with skepticism.

Perhaps the answer, as it relates to work, lies in understanding the current state of our workplaces. According to Gallup's most recent "State of the Global Workplace" report, only 15 percent of employees worldwide are fully engaged in their work, leading to a staggering $7.8 trillion in lost productivity annually.[1] This disengagement crisis doesn't fall solely at the feet of human resources; it's a major financial concern impacting both shareholders and clients.

Employees often feel trapped in a relentless grind, losing passion for their work. Employers, fixated on short-term key performance indicators (KPIs), tend to apply quick fixes rather than addressing underlying issues. Why is this happening? The answer is both simple and complex: our traditional models of employment are failing to adapt to the evolving needs and expectations of the modern workforce.

The dissatisfaction prevalent among workers today extends beyond workplace fatigue. It reflects a deep-rooted systemic failure in how we structure work. To effectively adapt, organizations must do more than just modify existing structures (the Band-Aid fix, if you will); they need to reimagine the very foundation of work. This is critical when the stakes are so high and change is accelerating. Disengaged employees, who are currently leaving to seek better employment opportunities, may soon be drawn to entirely new realms of work, such as digital roles within the metaverse or freelancing as a bounty hunter for a decentralized autonomous organization (DAO). The options available for workers have never been so expansive, threatening the stability and attractiveness of standard employment models.

The present employment crisis calls for more than just superficial solutions. It's like trying to fit advanced components of a Tesla into a Model T. The foundational structure needs a complete overhaul. This is where emerging technologies like generative AI, web3, and decentralized systems come into play.

These technologies propose a fundamental shift in how we approach work. They empower employees with greater agency, opportunity, creativity, and collaboration within new work structures where both individuals and organizations can thrive. For example, GitLab, a leader in the all-remote work sphere, is revolutionizing work structures by flattening traditional corporate hierarchies and enriching a culture of open information.[2] Its CEO attributes its success to a robust practice of documenting employee activities and methodologies. This approach encourages collaboration, as remote workers can access and edit documents that allow them to share feedback and insights across various company processes. GitLab's shift to remote work has prompted a significant change in its work ethos, emphasizing results over hours worked.

Eightfold.ai is leading the charge in revolutionizing traditional employment models using AI and data.[3] Its approach focuses on evaluating an individual's potential for job success by analyzing skills and experience, moving away from traditional reliance on university degrees and personal networks. Proponents of this technology argue that by leveraging data and sophisticated algorithms, the hiring process can become more equitable and diverse. Eightfold's AI-driven talent platform illuminates the path to hiring and nurturing individuals to their fullest potential.

And then there's gaming platform The Sandbox, which is pioneering a range of job opportunities in the metaverse, spanning various engagement levels, including full-time, part-time, and fractional positions.[4] This spectrum encompasses a diverse array of roles, from casual to competitive gaming, content creation, and mentorship, to the ownership and management of virtual real estate and goods. Additionally, roles in web3 education, event planning, and digital influencing are emerging. In a thriving digital economy facilitated by blockchain technology, there are so many opportunities to work in ways that were previously unimaginable.

As we contemplate the future of work, shaped by technological advances, we recognize that predicting exact outcomes should be left to the oracles. With backgrounds in innovation and human resources, however, we clearly see the imperative shift from a traditional employment model to a modern, decentralized framework. This transition is essential to provide workers with the needed flexibility and empower companies with the agility and innovation necessary for growth.

As we stand at the edge of a technological revolution, whether you're an avid fan or just starting to learn about these advancements, it's essential to recognize how these innovations can fundamentally transform the way we work and lead us toward a more productive and fulfilling work environment.

Unpacking the Significance

For today's leaders, understanding the new technologies, how they overlap, and the role they play in work should be an imperative for strategic vision and sustained relevance in the ever-evolving landscape of business. Corporate executives who dismiss or ignore this transformation risk falling behind competitors who adapt, innovate, and capitalize on the opportunities presented by these disruptive technologies. What's more, they lose out on key talent and major players in the industry who are looking for different work experiences that go beyond the traditional office setting.

The following are just some of what these disruptive technologies offer:

Competitive advantage and innovation. Whether harnessing the power of AI for operational efficiency or adopting decentralized work models for flexibility, innovation becomes a cornerstone of success. Those who lead the way in embracing disruptive technologies position their organizations as industry innovators, attracting top talent and gaining a strategic edge.

Optimized operational efficiency. The integration of automation, AI, and decentralized work models optimize operational processes, streamline workflows, and reduce costs.

Agile and improved work dynamics. Corporate executives who grasp the shift to decentralized work can proactively adapt their organizational structures, creating agile workplaces essential for maintaining a competitive edge.

Talent acquisition and retention. The workforce of the future seeks employers who embrace innovation and offer a dynamic work environment that is more human. By offering flexible work arrangements and staying at the forefront of technological advancements, organizations become magnets for skilled professionals.

Global impact and market expansion. Embracing disruptive technologies opens avenues for global impact and market expansion. Decentralized work models facilitated by blockchain and web3 technologies allow organizations to tap into a global talent pool and expand their market reach.

A future-proofed organization. Corporate executives who proactively engage with this transformation and invest in future-ready strategies are better equipped to navigate the uncertainties ahead.

This era stands out in the evolution of work due to the unique convergence of technological innovation, where AI is replacing cognitive work, and a concurrent societal readiness for transformative change. Diverging from past revolutions that relied on slightly manipulating systems, the ascent of decentralized work ecosystems, propelled by advancements in AI and blockchain technologies, is fundamentally reshaping our perspective on work.

About the Book

The inception of *Employment Is Dead* traces back to a moment of serendipity in September 2022. Josh Drean, a startup adviser from Harvard's Innovation Lab, was drawn to the WAGMAS web3 summit—a cutting-edge event cofounded by Silicon Valley serial entrepreneur and *New York Times* bestselling author Deborah Perry Piscione and renowned Silicon Valley technologist and web3 entrepreneur Val Bercovici—and submitted a speaker application. Among the flurry of applications, Deborah immediately recognized the importance of addressing the future of work as distinct and critical subjects. She was captivated by the intriguing premise of Josh's speech on the intersection of technology and new work models: "Emerging technologies will bring about a work revolution like never before in history. It will provide a new vessel and reality of work made possible by artificial intelligence, blockchain technology, cryptonetworks, and smart contracts... and how this new landscape will forever change our relationship with work."

Deborah decided to reach out directly to Josh, and taking a leap of faith, she asked, "Are you interested in writing a book together?" To her delight, Josh responded with a chuckle and an enthusiastic "Yes... yes, I am."

Now, as coauthors and cofounders of the Work3 Institute, an advisory firm that marries emerging technologies with workforce strategies, we research, advise, and lead on a topic that keeps us up at night: How do businesses capitalize on disruptive technologies to reimagine work to be more engaging, more human, and more democratized?[5] Bringing Josh's background in talent management, people analytics, and workforce experience and Deborah's in the Silicon Valley innovation ecosystem, we now help organizations adopt Work3 models for employment, giving leaders insights into the future of their labor force and how workers can supercharge their efforts by augmenting their cognitive functions and increasing company productivity and innovation, consequently, growing the pie to increase revenue.

In the chapters ahead, we'll explore these disruptive technologies on a quest to reimagine what work will look like as these technologies collide with workplace needs. From examples of companies on the leading edge to interviews with pioneers in the field, we will explore the myriad of ways these technologies are poised to redefine "employment" as we know it. And most critically, we will investigate how leaders can navigate these changes by adapting their management philosophies and operational structures to serve an increasingly agile and empowered workforce.

This book does not attempt to predict the future of the labor force or the exact future of work, as no one knows exactly where the lines will be drawn; however, the opportunities of decentralization are clear and the cases are compelling. Many different technologies are being tried and tested—some will fail, others will have moderate success, and yet some, it is guaranteed, will hit a grand slam. Just as today's tech giants were born out of the dot-com era, the new generation of innovative tech companies are working to establish their legacy today.

We hope this book will spark curiosity, help you get up to speed on today's disruptive technologies, and provide new insight into how the technologies are impacting the future of work. (To this end, we've included a quick rundown of key terms at the end of this introduction as a reference point if you need it as you read through the book.) While we champion the benefits of decentralization and the potential for a more human-centric work environment, we also address the challenges of ensuring that the human element remains at the forefront of organizational change.

For leaders, the risk here is not just falling behind on where the world is heading—it's about becoming irrelevant. You face the challenge of guiding your organizations through uncharted waters, all the while ensuring that the human element, often vulnerable in times of rapid technological adoption, is not lost. While these disruptive technologies promote a culture of autonomy and collective responsibility, they also call for a strong governance structure rooted in ethical practices. Balancing these elements is critical. Leaders who do this successfully will set the stage for a more resilient, adaptive, and ethically grounded workforce, providing humans with a work environment that is . . . well, more human.

Terms to Know

Artificial intelligence (AI): A set of computer systems or programs that can perform tasks that typically require human intelligence, such as problem-solving, learning, perception, reasoning, and language understanding.

Augmented reality (AR): An interactive experience that occurs by superimposing a computer-generated image on a user's view of the real world, thus providing a composite view.

Blockchain: An advanced database mechanism that allows transparent information sharing within a business network by storing data in blocks that are linked together in a digital chain.

Decentralization: The distribution of power, authority, and decision-making away from a central authority or governing body in an organization. Decentralization strives to satisfy the varying requirements for participation, independence, and status by promoting a spirit of group cohesiveness and spirit.

Decentralized applications (dApps): Applications that can operate autonomously through the use of smart contracts that run on the blockchain or other distributed ledger system.

Decentralized autonomous organization (DAO): An organization managed in whole or in part by decentralized computer programs, with voting and finances handled through a blockchain. DAOs are member-owned communities without centralized leadership.

Generative artificial intelligence (gen AI): AI that is capable of generating text, images, videos, or other data using generative models, often in response to prompts. Generative AI models learn the patterns and structure of their input training data and then generate new data that has similar characteristics.

Global talent marketplace: A digital talent platform driven by AI that allows access to the global talent pool and work with contractors from anywhere in the world.

Immersive internet: A version of the online world that incorporates advanced technologies to enhance user engagement and blur the line between the user's physical reality and the digital environment.

Intellectual property (IP): The legal framework that protects creative and innovative works. From AI algorithms to blockchain-based decentralized applications, the ownership, sharing, and protection of intellectual creations form the foundation on which these innovations stand.

Machine learning: The use and development of computer systems that are able to learn and adapt without following explicit instructions by using algorithms and statistical models to analyze and draw inferences from patterns in data.

The metaverse: A network of shared, immersive virtual worlds in which users can interact with a computer-generated environment and other users.

Non-fungible tokens (NFTs): A unique digital identifier that is recorded on a blockchain and is used to certify ownership and authenticity. It cannot be copied, substituted, or subdivided. The ownership of an NFT is recorded in the blockchain and can be transferred by the owner, allowing NFTs to be sold and traded.

Smart contracts: A self-executing contract with the terms of the agreement directly written into the code.

Web3: A concept used to describe the next generation of the internet, in which most users will be connected via a decentralized network and have access to their own data.

Virtual reality (VR): A completely immersive experience that replaces a real-life environment with a simulated virtual world.

PART I

The Work3 Revolution

CHAPTER 1

Welcome to Work3

Employment is dead.

It may seem alarming to hear such a claim, but in today's world of advancing and disruptive technologies, traditional employment is less and less applicable; in fact, it could be holding you back from attaining the business results you seek. Employees' preferences for work are dramatically shifting. They aren't thrilled to work in an office anymore. Many are questioning the standard nine-to-five hustle that has been so prevalent for a century. They want to leverage their skills, interests, passions, and backgrounds to do meaningful work, rather than complete monotonous tasks handed down from above. And this is just the tip of the iceberg.

Burnout is now a legitimate medical diagnosis according to the World Health Organization.[1] Quiet quitting, or doing the minimum requirements of one's job, has been adapted by even the most talented workers, and disengagement is at its highest point since Gallup started tracking it in 2000.[2] And it might be confusing to see the discrepancy between reading headlines saying employees are being lazy and reviewing employee sentiment about the discontent they experience in their daily work.

It's not that employees don't want to work. Most yearn to productively contribute to society. They just experience a disconnect from their daily grind to the greater good. The postpandemic workforce wants to work differently in a way that serves their interests as well, and if leaders aren't paying attention to these preferences and providing a better work experience, they'll soon find themselves unable to compete for top talent.

Let's distinguish between "employment" and "work," two terms commonly used as synonyms that are now being redefined. (See the sidebar "Employment versus Work.") The historical narrative of work includes a familiar scenario of traditional employment reminiscent of the industrial era. This model involves clocking in and out, commuting to an office, and functioning within the confines of set processes and rules, as defined by a manager. It's an archetype characterized by standardized work hours and hierarchical organizational structures. As we progressed into the twentieth century, significant developments occurred, from the introduction of labor laws to the rise of workers' rights movements. These changes gave shape to the conventional nine-to-five workday, a concept that has defined modern employment for decades.

But now, the conventional work structures of employment that have long governed our professional lives are being upended. Disruptive technologies are rewriting the social contract that has underpinned our organizations for more than a century. In *Workforce Ecosystems,* the authors share research that implies that the composition and boundaries of the workforce have already changed significantly and that formerly accepted management styles are ill-suited for workforces that

Employment versus Work

Employment: A contractual agreement between an employer and an employee for certain services, where compensation, in return, is a salary or hourly wage.

Work: Activities, tasks, or projects that an individual engages in to earn income, pursue passions, or contribute skills on a flexible, short-term, or long-term basis.

span internal and external organizational boundaries: "Historically, when we consider *'work,'* we think of processes along with a focus on consistency and efficiency. For centuries, standard operating procedures (SOPs) have dominated the landscape of those who study work processes. Today, from factory floors to software development, we see a shift to project-based work and an emphasis on outcomes."[3]

In essence, we're standing at the cusp of a transformative shift in how work is conceptualized and executed. This landscape offers a spectrum of options for individuals, allowing them to redefine employment, compensation, and governance on their own terms. As we navigate this transition, it is important to address the challenges and opportunities inherent to both traditional and digital work environments.

Why Things Are Different

What makes this era of work so interesting—and different—is the push toward an even more decentralized workforce. In essence, decentralization involves redistributing control, authority, and decision-making power away from a single centralized entity toward a more distributed and participatory framework.

Traditional organizations often entail rigid structures where decisions flow from upper management down to the workforce. In contrast, decentralized models promote a bottom-up approach, empowering employees at all levels to contribute ideas, make decisions, and influence the direction of projects.

As we look for a path forward at this intricate intersection of technology and societal evolution, recognize that the transformation in the world of work represents a shift in how we perceive our roles, obligations, and contributions in the professional sphere. The blurring of lines between the digital and physical realms is both a technological marvel and a cultural phenomenon that redefine our relationship with work. This symbiotic relationship calls for a nuanced understanding that requires organizations and individuals alike to adapt and innovate. In this era, where the digital and the societal are inseparable, embracing change becomes as much a strategic imperative as it is a cultural ethos essential for navigating the complexities of the modern professional landscape.

As we seek to understand the complex forces driving us toward a digital future of work, let's explore four major influences that are drastically reshaping employment as we know it.

A New Domain of Cognitive Work

The influence of AI and generative artificial intelligence (gen AI) will be profound, significantly shaping the future of work, transforming the labor market, and revolutionizing the roles of office professionals. While AI will certainly replace some repetitive tasks and jobs for individuals, the real promise lies in how AI and humans will work together.[4] A recent study by the International Labor Union, a part of the United Nations, has indicated that AI is more inclined to enhance job roles than eliminate them.[5] IBM CEO Arvind Krishna said during an interview with CNBC that AI is "absolutely not displacing—it's augmenting" white-collar jobs.[6]

Conversely, many business leaders think otherwise. In a personal interview with Edo Segal, the founder of Touchcast, a startup that reimagines the future of the generative web, has said: "We have never had a scenario where AI replaces the domain of a cognitive practice at this scale. Automation was originally intended to replace manual labor, but now it's possible to scale the automation of cognitive roles. We had narrow AI for narrow use cases, but not broad solutions like the ones emerging now that can replace entire professions like programmers, certain types of lawyers, and management consultants."[7] Recent research by Goldman Sachs supports Segal's claims and reveals that AI could replace the equivalent of three hundred million full-time jobs in the next fifteen years, impacting office jobs that were previously thought to be irreplaceable.[8]

With any technological evolution, there are winners and losers. According to the World Economic Forum's "Future of Jobs Report" on job growth in the next five years, "23 percent of jobs are expected to change by 2029, with sixty-nine million new jobs created," with the fastest-growing jobs in AI and machine learning specialists, among other web3 technology-related jobs.[9] Over the longer term, AI could eventually increase the total annual value of goods and services by 7 percent, if Goldman Sachs's AI growth projections are fully realized.[10] And, with AI investment forecasted to approach $200 billion globally by 2028, the technology could support humans in ways never before imagined.[11]

For now, the general consensus is that AI will have a positive impact on the future of work, making companies more profitable and productive. But this shift is also coming at a time when the traditional business model will be spun on its head by other disruptive technologies.

New Business Models

The emerging decentralized work model hinges on the distribution of authority and tasks, which promises to make work more agile and responsive to individual needs and collective goals. These technologies, often referred to as web3, form the bedrock of a groundbreaking shift in what's possible and how new work models will bring greater agility to companies seeking a competitive advantage. Web3 is introducing a range of novel business models, thanks to technologies such as blockchain, decentralized protocols, digital systems that operate without a central authority, and user ownership of data.

The gig economy, for example, is evolving into a global talent marketplace, where individual, independent workers—formerly known as employees—will have more power and control over their earnings and livelihoods. With effective implementation, web3 technology has the potential to address various business challenges and employee pain points. For instance, it can eliminate the need for middlemen or managers, allowing workers to engage directly with clients or customers. The use of smart contracts on the blockchain can automate and guarantee fair compensation to reduce conflicts and ensure prompt payments. Contributors can truly own their digital creations like art, music, and project contributions through blockchain and NFTs, giving them greater control and equitable remuneration.

Forward-thinking companies are starting to experiment with decentralized autonomous organizations (DAOs), where decision-making is distributed among contributors.[12] DAOs enable community-driven projects to recognize all participants involved as stakeholders to vote on proposals, investments, and governance matters. DAOs are like a digital democracy where you're no longer an employee but a cocreator in the project with direct investment in its success.

Along with the rise of DAOs comes the transition of employees from mere cogs in the corporate machine to empowered contributors who have a tangible stake in their work. Members of a DAO can have a direct say in decision-making processes, from resource allocation to strategic direction. This not only democratizes the workplace but also allows employees to retain much more of the value they generate.

In this system, every task completed, every idea contributed, and every project led can be directly attributed to an individual, who can then be fairly compensated and recognized. Unlike traditional setups where the organization harvests the fruits of your labor, DAOs ensure that value flows back to the people who create it. In essence, DAOs facilitate a culture of verifiable, transparent, and equitable ownership, thereby redefining what it means to be truly engaged in one's work.

And who is best poised to evangelize and utilize these web3 applications? Gen Z—the new wave of workers already well versed in digital technologies, decentralized systems, and a culture of innovation and social awareness.

A New Generation of Workers

Members of the new generation have called into question the traditional pathways to career success because they are experiencing the effects of its unraveling. College is more expensive than it has ever been, with no guarantee of employment after completion. They are struggling to afford the cost of living on one income, and a majority are forced to chase a side hustle. They've seen generations before them hustle and grind for forty years with the promise of a decent retirement—and even that model is slipping away. They'd rather prioritize living now while they are young enough to enjoy it. Thus they are pushing back on the traditional corporate narrative that they have to work to be happy.

In fact, they believe the opposite is true. Happiness isn't what you do; it's who you choose to be. And they are choosing a different path.

The term "youthquake," originally coined by *Vogue* magazine in the 1960s to describe the era's fashion and cultural shifts, has made a comeback to embody Gen Z's impact in the workplace for two key reasons: its dominating population and its innate digital fluency.[13] Approximately 52 percent of the global population is under age thirty, according to the US Census Bureau, and the members of this digitally native generation have had devices in their hands since they were toddlers, which has profoundly shaped their values, interests, and worldview.[14] Gen Zers often blend reality with the digital realm, sometimes even preferring to live, create, and work in the latter.

Consider Roblox, the gaming platform that allows users to play hundreds of user-created games. It has amassed a staggering 66.1 million daily users who actively engage in buying, selling, designing, and innovating within its virtual universe.[15] Many of these users, who have been actively playing open-source games for almost two decades, already believe they have a viable job because of the value they produce to earn the virtual currency awarded in these games, which they can even exchange for real money on the Roblox Developer Exchange Program (DevEx).

According to a recent EarthWeb poll, 75 percent of kids aged six to seventeen now aspire to be YouTubers, rather than traditional professionals such as doctors or firefighters.[16] This trend highlights the rising impact of the $250 billion influencer economy, where creative freedom often outweighs the appeal of traditional corporate jobs.[17] As we look ahead, it's crucial to consider what work will look like for a generation that has come of age with AI, blockchain, and decentralization.

A few trends have emerged from the Gen Z–led companies we've interviewed; for example, they hold "Mindfulness Fridays" for deep focus without meetings and encourage quarterly "heart checks" to see how direct reports are feeling about their workload and pay. They mandate a "slump hour" when employees are not allowed to schedule meetings immediately after lunch. They use emoji in abundance and curse freely on Slack. They can head to the gym in the afternoon and run some errands, knowing they'll be working into the night.

In short, Gen Zers don't want the traditional corporate nine-to-five. They are fluidly blending work and life together. They work to live rather than live to work. It's this youthquake that brings a fresh, unapologetically critical perspective to how work should be organized, compensated, and valued.

A New Perception of Work

Companies with flexible remote work policies outperform on revenue growth by 16 percent, according to a 2024 report released by Scoop in conjunction with BCG.[18] Another report by Harvard Business School suggested that remote workers were, on average, 4.4 percent more productive than their in-office counterparts due to quieter work environments, fewer interruptions from colleagues, and the ability to structure the workday to suit individuals when they are most productive.[19]

What's more, the gig economy—freelancing, temporary contracts, and project-based work—is becoming more prevalent, offering individuals greater flexibility in choosing their engagements and in choosing when, where, and how they want to make a sustainable living. Perhaps most importantly, we've also seen a notable mindset shift in terms of what people are willing to tolerate in their work lives going forward, as they place more emphasis on their well-being and purpose, in addition to environmental and social considerations for the world.

The next iteration of workplaces will place an increased emphasis on employee well-being, mental health, and sense of purpose. We can't divorce workplaces from the responsibilities of environmental sustainability and social responsibility. Businesses now understand that having motivated and content employees significantly boosts productivity and sparks innovation.

Today, there are established work practices that were once unconventional but are widely accepted. For instance, the use of Zoom meetings or Slack for remote collaboration, the flexibility to work from home during one's most productive hours, and even relaxed dress codes have all been accepted, despite the fact that they were resisted before the Covid-19 pandemic. But beyond breaking down geographical and temporal barriers is the dismantling of the age-old hierarchies that have governed the workplace. The result is a more equitable, inclusive, and dynamic work environment where value is cocreated in real time, globally. This isn't an extension of the old ways, but a comprehensive reimagining that promises to redefine how and why we work.

Work3

Understanding all these changes means accepting a new way of working. We define the next iteration of work as Work3, which encapsulates the dynamic evolution of how individuals, organizations, and industries approach and execute work tasks and activities. In the contemporary landscape, characterized by advancing technology and shifting societal trends, Work3 invites the following significant transformation:

Technological advancements. Work3 is closely tied to the integration of cutting-edge technologies such as AI, automation, machine learning, and advanced data analytics. These innovations serve to streamline processes, enhance efficiency, and facilitate the completion of tasks that were once manual or time-intensive.

Remote and flexible work. Work3 extends beyond traditional office settings, embracing the prevalence of remote work and flexible arrangements. Recognizing the importance of work-life balance, this iteration allows individuals to work from diverse locations and tailor their schedules to better align with personal needs.

Collaboration and connectivity. Emphasizing collaboration across geographical boundaries, Work3 relies on virtual teams, online platforms, and digital communication tools. These elements foster seamless collaboration among individuals from different parts of the world, contributing to a globalized approach to work.

Skills and learning. Given the rapid pace of technological evolution, Work3 places a strong emphasis on continuous learning and upskilling. Workers are required to adapt and acquire new skills to stay relevant in their respective fields as job requirements undergo dynamic changes.

Gig economy and freelancing. The ascent of the gig economy has reshaped work dynamics, with freelancing, temporary contracts, and project-based work gaining prominence. This trend offers individuals increased flexibility in selecting and engaging in various professional opportunities.

Well-being and purpose. Work3 underscores the significance of employee well-being, mental health, and a sense of purpose. Organizations acknowledge that motivated and fulfilled employees positively impact productivity and innovation.

Environmental and social responsibility. In alignment with broader societal trends, Work3 incorporates considerations of environmental sustainability and social responsibility into business practices. Organizations are increasingly mindful of the impact of work on the planet and society, reflecting a growing awareness of these crucial dimensions.

The concept of Work3 emerges at this juncture as an innovative framework that fully leverages the advantages of disruptive technologies and the empowerment of humans. It represents an evolutionary leap of merging the best elements from both traditional and decentralized work models. Work3 redefines how we work by transforming how we collaborate, make decisions, and distribute value.

Overall, the next iteration of work is characterized by its adaptability, integration of technology, flexibility, emphasis on collaboration, and a holistic approach that considers the well-being of individuals along with the broader societal impact of work. Now let's dig in a little further to decentralization and its potential to revolutionize employment as we know it.

Decentralized Workforce Ecosystems

On October 31, 2008, the Satoshi Nakamoto white paper, "Bitcoin: A Peer-to-Peer Electronic Cash System," was released. Though the thesis of the paper didn't explicitly state societal objectives, many interpret the white paper as a call to democratize finance. The goal of the peer-to-peer system aimed to remove intermediaries in a banking transaction and bring financial inclusion to the two billion unbanked individuals and two hundred million small businesses globally, which forever challenged the status quo.[20]

Nakamoto's white paper highlights the power of decentralized groups working toward a common goal, and inherently, the need for a new form of governance at work that democratizes the decision-making process and improves society. By championing a meritocratic ethos and valuing individuals based on their skills and contributions rather than conforming to traditional hierarchical structures, the organization begins to organically grow toward a brighter future without all of the red tape and communication blockage.

This shift challenges traditional top-down hierarchies by granting individuals greater autonomy and enabling collective decision-making. From the realms of blockchain-driven financial systems to global talent marketplaces and even social media platforms, decentralization has been gaining momentum due to its potential to advance transparency and innovation (see figure 1-1).

This model promotes a healthy work-life balance while tapping into a global talent pool, which can tap into deeper wells of innovation and diverse perspectives. However, this demands astute leadership, adept management of virtual interactions, and a culture that thrives on trust and accountability. In this fast-paced digital world, here's something you can't ignore: decentralization holds the key to an agile and innovative workforce. It requires letting go of traditional power hierarchies for the sake of growth, and that's just how modern companies will win. Remember Mark Zuckerberg's famous motto, "Move fast and break things"? Now it's "move faster with stable infrastructure."

FIGURE 1-1

Decentralization visualized

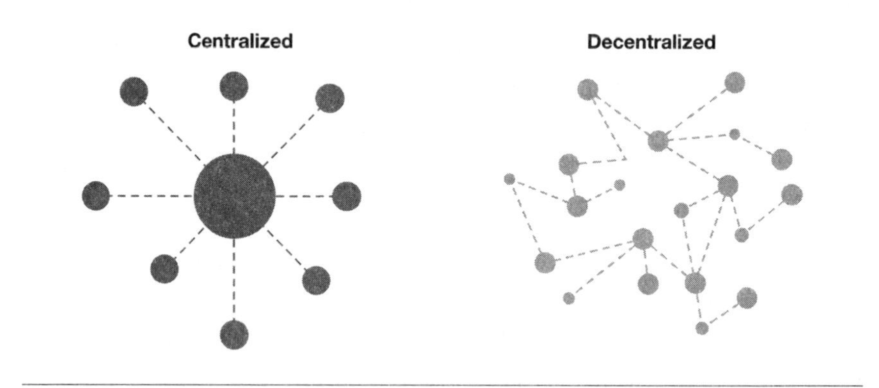

A Bifurcated Work Economy

Not every industry is well-suited for decentralization. While creative fields like media production, software development, and design appear particularly adaptable to decentralized models, as they naturally thrive on diverse inputs and benefit from flexible work structures, some sectors, such as utilities, health care, and defense, benefit from centralized systems due to the critical importance of streamlined decision-making, regulatory compliance, and secure data management.

But that's not to say change isn't happening. Timing is crucial in all this. Industries undergoing rapid technological advances may soon find decentralization advantageous, while those deeply entrenched in traditional workflows may need more time to adapt. Both the nature of the industry and the timing of technological adoption are pivotal factors in determining the effectiveness of decentralization.

Thus, the development of a bifurcated work economy, where traditional employment coexists with a decentralized web3 pipeline of projects, represents a significant shift in how individuals engage in work and contribute to the global economy. In this dual system, some industries will continue to rely on traditional employment as a foundation, while others may move toward a hybrid model that combines full-time employees with contractors and other contributors. While many people may still opt for the perceived stability of salaried positions in established organizations like corporations, government bodies, or nonprofits, others will be drawn to the flexibility that comes with working for multiple DAOs, allowing them to diversify their income streams to suit their lifestyles and skill sets. Some workers may prefer or require a standard paycheck, while others are eager for digital tokens or other cryptocurrency.

This new breed of work is dynamic and specialized; workers can focus on niches where they excel, enhancing both their market value and the efficiency of the projects they engage in. The flexibility intrinsic to web3 work can translate to improved work-life balance and job satisfaction. And workplaces should adapt because of it.

As these two economies—the traditional and the web3—continue to coexist and interact, leaders will need to navigate both worlds simultaneously in order to tap into the vast talent resources available.

Work without Jobs

Since graduating from the University of Michigan in 2021, Chase Chapman has opted out of the traditional employment route. Instead of taking a corporate job, with all its security and predictability, she has jumped in with both feet to become a contributing member of several DAOs—including Index, Forefront, RabbitHole, and Orca—and invested in several web3 side projects. She gets a rush out of working across multiple organizations simultaneously, hopping in and out of projects that excite her, and takes full advantage of the autonomy and flexibility that comes from collaborating within these work environments.

"We used to think that our jobs were these things that should give us fulfillment professionally and completely," Chapman said in an interview for CoinDesk's Future of Work Week. "Now you have a much more fluid way of thinking about what it means to do different types of work and engage across several different organizations." In a widely circulated article, Ben Schecter argues that these work models are better than traditional companies at orchestrating projects by providing better alignment between organizational goals and individual benefits and allowing people to mix and match an array of income streams and ownership returns.[21]

Steve Glaveski paints the picture of these potential work models: "Instead of having one employer and a 40-hour workweek, we'll likely contribute several hours a week to several DAOs. This technology-centric nature will result in rudimentary, algorithmic work being automated, freeing contributors up to be the most creative and useful versions of themselves and allowing them to spend more time on high-value activities—the type that stimulate the flow state—and less time on monotonous, shallow tasks."[22]

Indeed, the combination of these emerging technologies has the potential to radically transform operating models and the labor market. Ravin Jesuthasan, *Wall Street Journal* bestselling author, explains in his book, *Work Without Jobs*, "New types of businesses will emerge that will look more like cooperatives and less than corporations, significantly reducing agency costs. In such decentralized organizations, leadership will rely on soft power and empathy, using culture and shared values to align the interests of disparate stakeholders to a common mission and purpose."[23]

As we move forward, keep in mind that a majority of the technologies we'll examine still have to prove themselves as bedrocks of the new economy; nothing is set in stone. But the pace at which they've been attracting talent, capital, and innovation energy is astounding. Mainstream proliferation could happen sooner rather than later. Early adopters will find themselves riding the wave toward a new era of work and distributed prosperity, while the laggards will miss out on the opportunity to be a part of the revolution. So, while the outdated concept of employment is being loaded onto the tow truck and heads to the scrapyard, let's explore the shiny new options that will drive forward a more democratic and equitable future of work.

CONVERSATION STARTERS

1. How can you bridge the gap between traditional employment structures and the emerging decentralized work ecosystem to harness the best of both worlds?
2. How can you redefine leadership roles and responsibilities in a decentralized work environment where traditional hierarchies are disrupted?
3. Considering Gen Z's digital fluency and unique expectations for the work environment, what approaches can businesses take to prepare for and integrate this new generation into the workforce?
4. What measures can your organization take to ensure that the empowerment offered by DAOs and web3 technologies leads to a tangible ROI?
5. Regarding the bifurcated work economy, how can companies in sectors less suited to decentralization innovate to stay competitive?

CHAPTER 2

Ten Principles of Decentralized Work

Anne Shoemaker was just fourteen years old when she designed her first game in Roblox, a metaverse gaming platform launched in 2006 by David Baszucki and Erik Cassel. Anne's game was called *Mermaid Life*, and it quickly became incredibly popular, so much so that she won the 2022 Roblox Innovation Award. When asked on a CNBC interview how much she made that first year, she shocked the host by announcing that she made $500,000.[1] She went on to build other successful games, founded Fullflower Studio and was spotlighted on Forbes 30 under 30.

With an impressive 77.7 million daily users, Roblox has emerged as a dominant force, captivating the time and attention of Gen Zers and Gen Alpha. The platform's user-generated content model and immersive virtual experiences have revolutionized digital interactions, nurturing a dynamic community where collaboration, networking, and entrepreneurship flourish. By engaging in this vibrant ecosystem, young creators like Anne are not only expanding their networks and forging valuable relationships but also gaining priceless insights into transforming their passions into lucrative ventures.

The financial rewards of developing successful games on Roblox are substantial, with developers earning a record $741 million across twelve thousand developers in 2023 alone.[2] While Anne's $500,000 earnings may be an outlier, the average game developer can still earn an impressive $61,000 a year—a significant salary by any standard. This raises the question: Why would any young person choose to flip burgers for minimum wage at McDonald's when they could work from home and earn a living by creating engaging content in a virtual world like an IKEA store in Roblox?

Roblox and other major gaming platforms have paved the way for a new era of work, reshaping the expectations of an entire generation. Gen Z values flexibility, community, ownership over their work, transparency, and autonomy—qualities that are inherent to the Roblox ecosystem. As this generation enters the workforce, employers must adapt to meet these expectations or risk losing out on top talent (see figure 2-1).

To navigate this evolving landscape, companies should embrace the following ten principles that address the deeper structural issues of employment and provide a path forward for everyone, not just shareholders or executives. By embracing the principles that have made Roblox a global phenomenon, we can unlock the full potential of our workforce and create a future where work is not just a means to an end but a source of fulfillment, innovation, and limitless opportunity. The choice is clear: embrace the future of work or risk becoming as obsolete as a traditional minimum-wage job at McDonald's in the eyes of Gen Z.

With that, let's jump in.

FIGURE 2-1

The ten operating principles of Work3

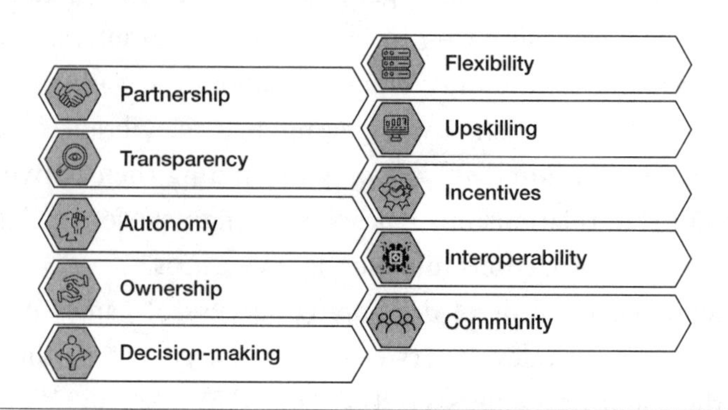

1. Partnership

Traditionally, employment has been seen as a contractual obligation: employee completes X; employer pays Y. Straightforward. Simple. But this transactional view is limiting. It reduces the richness of human potential and interaction to mere line items, relegating the scope of creativity, innovation, and shared achievement to the fine print. Contracts lay down the rules, but they often don't leave much room for vision or for growing together toward mutual goals.

It's time to drop the mentality of "work is a contract," and embrace the new mindset of "work is a partnership." This human-centric model will fundamentally change the mechanics of your work environment. Employers should be looking beyond resources to fulfill tasks and start seeking partners to innovate, steer, and grow with them. Employees, too, are seeking more than just a paycheck; they're looking for opportunities to cocreate value, influence, and contribute in meaningful ways.

Partnerships, whether between an employee and an employer or even two collaborating organizations, are living, breathing relationships, nurtured by shared objectives, reciprocal investment, and a common vision for what could be. They adapt and evolve over time, built on a foundation of mutual trust and respect. Just as you wouldn't enter a business partnership with the sole objective of extracting value from the other party, equitable work partnerships function on the premise that both parties are equally invested in each other's growth and success.

This represents a shift away from the conventional employer-employee dynamic—that age-old pyramid model—to a more symmetrical partnership-driven interaction. It's a two-way street where everyone involved stands to gain something valuable from the relationship.

Consider the ecosystem of talent marketplaces like Upwork or Freelancer. These platforms are essentially democratizing the labor market. No longer do workers need to fit into a single, predefined role in a monolithic organization. Instead, they can pick and choose projects that resonate with their skill set, ambitions, and even life circumstances. Meanwhile, companies can tap into specialized skills on a project-to-project basis. It's a win-win situation and an excellent example of equitable partnerships in action.

Equitable partnerships are no longer the "nice to have" of forward-thinking companies; they're quickly becoming the "must-have" for any enterprise eyeing sustainable success.

2. Transparency

Transparency is the cornerstone of Work3 and is the most highly sought-after ideal in forward-thinking companies. Transparency may initially present challenges for organizations seeking to implement it, largely due to conventional practices that prioritize the close safeguarding of information. Historically, data and strategic insights were confined to a limited group within the top of the corporate hierarchy, primarily to maintain existing power structures and managerial control. This approach relegated the broader workforce to a need-to-know status. However, contemporary business landscapes have evolved, prompting a significant reevaluation of these traditional practices.

Today's emphasis on transparency transcends aspirational values by serving as an operational imperative, especially in decentralized business environments. This concept extends beyond traditional approaches like open-door policies or frequent information-sharing meetings. Rather, it embodies a revolutionary level of openness facilitated by the capabilities inherent in blockchain and other web3 technologies.

Envision an organization where every team member has real-time access to critical financial data, project statuses, and customer feedback, or smart contracts that render previously opaque processes transparent and automated. Such scenarios are now operational in enterprises that have embraced decentralization. Blockchain technology ensures the immutability and public verifiability of records, thereby establishing an unprecedented level of trust and transparency.

Recall when Elon Musk publicly released all of Tesla's electric car patents. That's transparency with a capital *T*, aimed not just at enlightening his team but the world at large. That move rippled through the automotive industry, and while it seemed like a loss of competitive edge, it actually strengthened Tesla's position as a leader in sustainable technology. Transparency can indeed be a strategic advantage.

This culture of openness levels the playing field where everyone feels empowered to share ideas and even constructive criticism. Tools like DAOs take this to the next level by allowing transparent voting mechanisms where every contributor gets a say in matters typically reserved for the C-suite in traditional organizational structures. This results in a deeply ingrained sense of collective ownership and accountability, which can supercharge motivation and performance.

According to a study by Label Insight, 94 percent of consumers are likely to be loyal to a brand that offers complete transparency.[3] The same logic applies to the employment arena. When people know what's going on, they're more likely to stick around and give their best. Transparency underpins the entire ecosystem.

3. Autonomy

No more agonizing wait for approvals, no more stifling micromanagement. Instead, frontline workers and middle managers, armed with unique perspectives and empowered by trust, become decision-making dynamos. Sounds radical, right? A Gallup poll throws down some hard numbers: organizations embracing autonomy witness a whopping 21 percent jump in profitability and a 17 percent surge in productivity—proof that this isn't just feel-good fluff.[4]

So, how do you tap into autonomy without descending into organizational chaos? The secret lies in striking a delicate balance, starting with a rock-solid foundation of shared values and goals. Think of it as everyone singing from the same sheet music, understanding the "why" behind the "what." Once everyone is aligned, the "how" becomes a fertile ground for innovation, no longer constrained by rigid processes.

Imagine a marketing team focused on boosting brand visibility by 30 percent within the next quarter. Armed with transparency and trust, a team member in Barcelona might dive into influencer partnerships, while their Singapore counterpart explores podcast advertising. There's no need for bureaucratic hurdles or endless approvals; they simply get cracking, knowing their individual efforts contribute to the shared goal.

4. Ownership

Rather than relegated to the confines of being a *worker*, individuals become true stakeholders in organizations. Ownership is a tangible share of the value one creates, signifying a fundamental shift in how we understand employment and compensation.

Ownership moves beyond the old structures where the fruits of labor mainly benefited a concentrated group of stakeholders or higher-ups. Instead, this principle assures that those contributing their skills, creativity, and time to a project have a legitimate claim on its success. It's as if you're not just an artist hired to paint a mural, but you're also given a share of the revenue generated from those who come to see it. This piece-of-the-pie approach ensures a more equitable distribution of value and financial gain across the entire ecosystem.

Many progressive organizations are already incorporating this principle into their DNA. Consider Gitcoin, a decentralized platform that financially empowers open-source developers. Not only do contributors get paid for their work, but they also receive governance tokens that allow them to vote on the direction of the platform, effectively making them co-owners. Or look at Uniswap, a decentralized finance (DeFi) platform, where liquidity providers are rewarded with transaction fees generated from the trading volume of the assets they help to support. In both cases, the contributors are granted ownership stakes that are directly linked to their level of contribution and the value they add to the platform.

The real beauty of ownership lies in the sense of empowerment and accountability it engenders. Employees—or rather, co-owners—become more engaged, committed, and inspired to contribute their best, knowing that they will share in both the risks and rewards of their collective endeavors. This leads to a more equitable and fulfilling future for all.

5. Decision-Making

Decentralized decision-making streamlines processes and eliminates lengthy approval chains. Teams closer to the customer or project have the context and information needed to make informed choices without waiting for layers of bureaucracy. This agility allows organizations to react quickly to changing market conditions, seize emerging opportunities, and navigate unforeseen challenges with greater nimbleness.

Imagine a product development team faced with a sudden shift in consumer preferences. In a traditional structure, it would be stuck waiting for approvals from higher-ups, potentially missing a crucial window of opportunity. In a decentralized model, it has the autonomy to analyze the situation, leverage its on-the-ground expertise, and make agile decisions, adjusting its campaign in real time. This empowers individuals to act on their knowledge and expertise, fostering a more responsive and effective organization.

When individuals have the authority to make decisions, they naturally feel a greater sense of ownership and accountability for their outcomes. This drives a culture of engagement and motivates employees to go the extra mile, knowing their contributions directly impact the success of their projects and the organization as a whole.

Decentralized decision-making thrives on a foundation of trust and transparency. Organizations must empower individuals with the necessary information, training, and support to make informed choices. This open and collaborative environment nurtures mutual trust and encourages people to share ideas and insights freely, leading to better decision-making across the organization. Of course, decentralizing decision-making is not without its challenges. Establishing clear guidelines and boundaries is crucial to ensure alignment with overall organizational goals. Additionally, organizations must invest in developing the skills and capabilities of people to make sound decisions effectively.

6. Flexibility

The traditional limitations of physical workspace and rigid schedules dissolve in the face of decentralized work, offering unparalleled flexibility across hours, locations, and work methods. This empowers employees to tailor their workflows to their individual preferences and needs, demonstrably contributing to increased work-life balance and overall well-being.

However, Work3's transformative approach surpasses mere convenience. It shatters the confines of the standard nine-to-five work track and unleashes a radical spectrum of flexibility unimaginable in conventional settings. Gone are the limitations of location, single-employer restrictions, and narrowly defined roles.

Imagine a skilled software developer whose passion extends to graphic design. In a web3-driven environment, their expertise expands beyond coding tasks for a single company. They can contribute their design talents to a DAO focused on digital art, while also joining an education technology project on another DAO. This allows them to diversify their income, potentially garnering equity-like tokens across these initiatives, while their skill set empowers multiple organizations toward shared goals.

Beyond remote work, Gartner's research shows an increase in organizations leveraging contingent or freelance workers.[5] This signals a shift toward mutual recognition of flexibility's benefits. Companies gain access to specialized skills on a project-by-project basis, while workers achieve desired work-life balance by contributing to multiple stimulating projects.

This level of flexibility forms the bedrock of decentralized work's future. It embodies the autonomy and freedom inherent in such environments. Fueled by the previously discussed shift toward trust and empowerment, flexibility becomes both a principle and a practice, fundamentally reshaping twenty-first-century employment frameworks.

7. Upskilling

Upskilling for the Work3 revolution is about unlocking individual and organizational potential where individual strengths come to the forefront. Gone are the days of rigid job descriptions and siloed expertise. Instead, decentralized teams thrive on leveraging the diverse skill sets and perspectives of each person.

Decentralized work dismantles the traditional one-size-fits-all approach to talent management. Instead, it recognizes the unique value each person brings to the table. This shift necessitates a focus on identifying and nurturing individual strengths and passions. By providing personalized learning opportunities and encouraging cross-functional collaboration, organizations can empower people to contribute their best selves, leading to increased engagement and overall performance.

This newfound emphasis on individual contribution demands a commitment to continuous learning. In a decentralized landscape, knowledge and adaptability become key differentiators. Workers must stay abreast of industry trends, embrace new technologies, and actively seek opportunities to develop their skill sets.

Disruptive technologies emerge at breakneck speed, and industry trends evolve rapidly. To thrive in this environment, individuals must cultivate a growth mindset, embracing continuous learning as a critical skill. This involves actively seeking out new knowledge, experimenting with emerging technologies, and staying connected with industry thought leaders. But upskilling efforts should go beyond technical skills and focus on developing soft skills like critical thinking, problem-solving, and communication. These skills allow people to navigate change, learn new things on the fly, and collaborate effectively in ever-evolving environments. After all, those who can adapt quickly and seamlessly are highly sought after.

While individual initiative is crucial, organizations play a vital role in developing a continuous learning culture. This means investing in learning and development programs, providing access to relevant resources, and encouraging knowledge sharing through mentorship and peer-to-peer learning initiatives. Moreover, fostering a growth mindset at the organizational level encourages experimentation, embracing failure as a learning opportunity, and ultimately propels the entire organization forward in the ever-changing landscape of decentralized work.

8. Incentives

The notion of incentives in a decentralized organization marks a significant departure from the traditional salary and benefits model that has historically characterized employment. In this evolving landscape, incentives are more specialized and customizable, encompassing a diverse array of rewards. These can include everything from token-based rewards to profit-sharing plans and proprietary digital assets. The unique feature of these incentives is that they are deliberately designed to align with both an individual's contributions and the overarching objectives of the organization.

The problem with conventional incentives like annual bonuses or promotions is their frequent disconnect from long-term organizational objectives and overall success. A Deloitte report points out that companies with robust incentive programs experience a 14 percent uptick in employee engagement compared with their counterparts without such programs.[6] In a decentralized setup, the bridge between what one contributes and what one receives is more transparent and often immediate. The upshot? Enhanced engagement, sparking innovation and driving overall productivity to new heights.

Profit-sharing models and equity arrangements also reflect this principle in a compelling way. Take employee stock ownership plans (ESOPs) as an example; these are becoming increasingly prevalent, especially among startups. ESOPs grant employees a real stake in the organization, aligning their personal success with the success of the company.

While ESOPs and other equity arrangements offer a strong foundation for aligning employee and organizational interests, they can sometimes be too broad to effectively incentivize specific behaviors or achievements. In a decentralized work environment, it becomes even more effective to tie incentives directly to projects, tasks, or milestones completed. This granularity in incentivization provides immediate feedback and rewards, enhancing motivation and accountability. By structuring incentives in this way, organizations can catalyze higher levels of engagement and productivity, as employees see a direct correlation between their contributions and their rewards.

So, as we venture further into the realm of decentralized work, it's imperative to think innovatively about incentivizing human capital. As we've said before, this is not about making tweaks or adjustments to the existing systems. It's a restructuring that places the holistic well-being and active participation of individuals at the very heart of the organization. By offering a more dynamic and responsive set of incentives, we create a win-win for everyone involved.

9. Interoperability

Use of technology and collaboration tools is a key principle. Decentralized work relies on digital platforms to facilitate communication, project management, and collaboration among geographically dispersed teams. Let's say that you're a designer in London collaborating with a developer in Tokyo on a groundbreaking project. Both of you use different tools, speak different languages (metaphorically and literally), and belong to separate organizations. How do you actually get things done? Interoperability.

Imagine working in perfect harmony with geographically dispersed colleagues, utilizing different tools without friction. Interoperability bridges these divides, ensuring smooth file sharing, information accessibility, and a collaborative experience regardless of platform preference. Gone are the days of struggling to understand data or navigate incompatible systems; now, collaboration becomes a seamless dance, even across borders and platforms.

Interoperability is the translator, the bridge builder, the secret handshake that makes everything work in this decentralized world. Without it, communication becomes a game of broken telephone, data gets stuck in silos, and collaboration becomes a nightmare. In the world of technology, interoperability refers to the ability of different systems, devices, or software to work together seamlessly, even if they were created by different developers or have different designs. Just like a translator helps people who speak different languages understand each other, interoperability acts as a "translator" between various technologies, allowing them to communicate and exchange information effectively.

Interoperability acts as a "bridge builder" by creating connections between distinct technologies, platforms, or systems, enabling them to interact and work together smoothly.

For example, imagine you have a smart home with devices from different manufacturers, such as a smart thermostat, smart lights, and a smart lock. Interoperability ensures that these devices can communicate with each other and with your smartphone app, allowing you to control them all from a single interface. In this case, interoperability is the "translator" that enables these devices to understand each other's language and the "bridge builder" that connects them, creating a unified, efficient smart home system. But interoperable platforms allow everyone to be on the same page, regardless of the app or platform they use. Data flows freely, progress is clear, and you can focus on what truly matters: creating something awesome.

Individual skills are in high demand, and as we've learned, people shouldn't be locked into just one platform or organization. Interoperability ensures that individuals can own their data and seamlessly move it between projects and platforms. Additionally, blockchain networks, the backbone of many Work3 initiatives, offer connections across the world. Without interoperable protocols, individuals would be isolated islands, unable to trade or communicate. Thankfully, interoperability breaks down these barriers, enabling secure and efficient transactions across diverse systems.

What's more, data silos and clunky integrations become relics of the past, offering speed and efficiency. Interoperability streamlines workflows by eliminating these roadblocks, allowing for efficient data exchange and seamless integration between tools. This translates to increased productivity, as teams can focus on achieving common goals rather than battling technical hurdles. The impact on innovation is amplified, as streamlined workflows foster a dynamic environment where creativity can flourish.

Interoperability acts as a key that unlocks cross-industry collaboration and knowledge sharing. With this newfound ability to seamlessly connect and exchange ideas, the potential for innovation explodes. Groundbreaking solutions can emerge when diverse perspectives and expertise from various industries converge, facilitated by interoperable systems. Interoperability fuels a future where collaboration supports innovation at an unprecedented scale.

10. Community

The future of work isn't just about technology, financial incentives, or even ownership; it's about the relationships we build and the communities we become part of. The idea of "going to work" is evolving from a solitary trek to the office into becoming an active participant in a broader, often global, ecosystem.

In a community-centric environment, contributors are not isolated islands of productivity but integral threads in a colorful tapestry of experiences and ideas. Each contributor brings their own unique value, and the collective becomes richer for it. Community allows for a pooling of resources and knowledge, increasing the speed of innovation and problem-solving. It turns the traditional model of competitive individualism on its head, advocating instead for cooperative success.

Discord is a perfect example of how community dynamics are being leveraged for productive workspaces. Originally designed for gamers, Discord has evolved into a platform where professionals, creators, and organizations build private communities to collaborate and communicate. In these servers, you'll find different channels dedicated to different projects, discussions, and even casual watercooler chat. Unlike traditional corporate communication tools, Discord is community-driven, providing a space where every member can contribute to the conversation, share resources, or ask for help. It's like a digital coworking space filled with experts in your field, just a click away, regardless of their geographical location.

Moreover, Discord has become instrumental in the development of community-led projects. DAOs, for instance, often use Discord as their main hub for communication and governance, allowing for a more direct line between decision-makers and community members. This approach paves the way for truly decentralized development, where ideas and revisions can flow freely, be debated, and be improved in real time by a diverse set of stakeholders.

The shift toward community is a rethinking of how work can be structured for the benefit of all involved. Through community, we're recognizing the multifaceted contributions that individuals can offer. It's not just about what you can do but also about who you are, what you value, and how those personal attributes add to the collective good. As we adapt to the future of work, the principle of community takes center stage, reminding us that human connection and collective action are vital. They serve as the social glue in the decentralized workspace, enriching our professional lives in ways that are both measurable and immeasurable. Work can be more than just a job—it can be an inclusive and caring community.

A Human-Centered Approach to Decentralized Work

As the decentralized future unfolds, critical questions emerge: How can we navigate its complexities while ensuring a human-centered approach? How do we apply these ten principles in a way that offers growth to our business but focuses on each individual contributor? The responsibility lies with each person in the organization—from mid-level managers driving transformation to frontline employees seeking a sustainable future.

First, we must reimagine leadership. Ditch the image of an isolated leader making all the decisions, which trickle down through the organization; instead, envision a collaborative network where leadership emerges from diverse perspectives and expertise throughout the company. Encourage cross-functional collaboration, empower teams to make informed decisions, and celebrate shared successes.

Second, cultivate a culture of learning. In Work3, the ability to learn and adapt becomes a critical differentiator. Encourage individuals to pursue continuous learning, experiment with new technologies, and actively seek opportunities to develop their skill sets. Think of a team that readily embraces an emerging technology, not with fear but as an opportunity to enhance its capabilities and tackle challenges with renewed creativity.

Finally, remember the importance of shared ownership and trust. Decentralization thrives on a foundation of transparency and mutual respect. Cultivate a culture of open communication, where individuals feel empowered to contribute their ideas and expertise. Imagine a transparent environment where data empowers decision-making, fostering a sense of ownership and accountability across the organization. By embracing these principles, we can ensure that Work3 is not just about efficiency and agility, but also about fostering a work environment that is engaging, fulfilling, and ultimately, human-centered.

CONVERSATION STARTERS

1. How can you communicate and reinforce Work3 principles throughout your organization for widespread adoption? What steps does your organization need to take to transition from a traditional hierarchical structure to a decentralized, more inclusive Work3-inspired model?
2. What opportunities exist for cross-functional collaboration to drive innovation? How can technology support the adoption of Work3 principles, especially for remote or distributed teams?
3. How can Work3 principles improve employee satisfaction, retention, and well-being for higher productivity? How can you promote continuous learning and skill development?
4. How can you be trained to effectively lead in a Work3 context, emphasizing autonomy and trust?

PART II

The Biggest Disruptors at Work, Explained

CHAPTER 3

The Rise of Generative AI and Human+

"We're at an exciting turning point where artificial intelligence [AI] is evolving from a tool to a teammate," said Dustin Moskovitz at Asana's Work Innovation Summit in San Francisco in the summer of 2024. As cofounder of Asana, a work management platform that helps teams manage, track, and automate work, Moskovitz emphasized that "the real promise emerges when we combine human intention, judgment, and empathy with the capabilities of AI." Moskovitz believes that we stand at a pivotal moment in history, when we should no longer view AI with skepticism, but rather as a transformative force that will revolutionize the way we work. "Imagine an AI teammate that is the most organized, knowledgeable, effective, and encouraging project manager you've ever worked with, helping you figure out the best way to plan and accomplish work, and even doing a lot of the work itself," Moskovitz explained. "By thoughtfully integrating AI into our workflows as a collaborative partner, teams can radically boost their productivity, impact, and scale. It's this synergy of humans and AI working together that will truly transform what we can achieve."

The current state of AI is truly revolutionary, with the potential to significantly enhance and improve our work lives. Many of us have already experienced the benefits of using AI tools in our daily tasks.

More than ever, AI and its subset, generative AI (gen AI), are transforming the way people work and reshaping the skills required in organizations. Gen AI autonomously creates novel, contextually coherent, and humanlike content across various domains, including text, images, music, and more, by leveraging deep learning models to replicate human creativity. Before diving into what gen AI means for the future of work, let's take a moment to understand its origins.

The Evolution of AI

In the summer of 1956, John McCarthy, Marvin Minsky, Nathaniel Rochester, and Claude Shannon organized the Dartmouth Workshop at Dartmouth College in Hanover, New Hampshire. These visionaries shared a common interest in exploring the possibilities of creating machines that could simulate human intelligence. During the workshop, participants discussed ambitious goals, envisioning machines capable of performing tasks such as language translation, problem-solving, learning, and pattern recognition.

The Dartmouth Workshop led to significant outcomes, including early AI programs like the Logic Theorist, which proved capable of solving mathematical theorems.[1] Another milestone occurred in 1957 with the development of the General Problem Solver, designed to address diverse problem-solving scenarios such as logical reasoning puzzles, mathematical proofs, and strategic planning in complex environments.[2] The General Problem Solver played a crucial role in the early history of AI and cognitive science. While the workshop is widely acknowledged as the birthplace of the term "artificial intelligence," the conceptual foundations of AI were gradually taking shape long before this landmark event.

Laying the Foundation

Alan Turing is widely regarded as a pioneer of AI. In 1936, he introduced the theoretical concept of machine intelligence and the potential for creating machines with humanlike thinking capabilities through his invention, the Turing machine. The Turing machine, a theoretical model of computation, simulates the logic of computer algorithms and remains a cornerstone in the theory of computation. But Turing's contributions to the field extended beyond just this groundbreaking concept. His wartime efforts at Bletchley Park played a pivotal role in breaking the German Enigma code during World War II, and, in 1950, Turing proposed the Turing test as a measure of a machine's intelligence, which has since become a fundamental concept in AI.

The Rise of Machine Learning and Deep Learning

But AI's historical roots don't end there. Geoffrey Hinton, born in 1947, was a central figure in AI's modern era.[3] After studying experimental psychology and obtaining his PhD in AI, Hinton and his colleagues popularized the backpropagation algorithm in the 1980s. This algorithm became fundamental to training artificial neural networks, marking a cornerstone in the field of deep learning, a subset of the broader concept of machine learning.

Machine learning is a branch of AI that focuses on enabling computers to learn and improve from experience without being explicitly programmed. In other words, it's a way of teaching machines to learn by themselves. Deep learning, on the other hand, is a specific method within the machine learning framework that involves neural networks with many layers (deep neural networks). These networks are inspired by the structure of the human brain and are designed to automatically learn and represent complex patterns and features from data. While machine learning encompasses the overarching idea of self-learning machines, deep learning emphasizes the use of complex neural network structures to achieve this goal.

The 1980s witnessed the rise of expert systems, rule-based programs crafted to emulate human decision-making. The concept behind expert systems involved encoding human expertise, often gathered from interviews with domain experts, into a set of rules and logical statements. The 1990s then marked a pivotal era in the evolution of AI, exemplified by significant milestones such as IBM's Deep Blue triumphing over chess champion Garry Kasparov in 1997. This momentous event signaled the increasing prowess of AI systems in mastering complex tasks previously thought exclusive to human intelligence.

In the first decade of the twenty-first century, Hinton's work on deep learning began to gain traction. In 2006, he cofounded the Neural Computation and Adaptive Perception program at the University of Toronto, which became a leading center for research on deep learning. In 2012, Hinton joined Google AI, where he continued to work on deep learning research.

Now known as the "Godfather of AI," Hinton and his work on deep learning have had a profound impact on the field of AI. Deep learning is now used in a wide range of applications, including image recognition, natural language processing (NLP)—which focuses on enabling computers to understand, interpret, and generate human language—and machine translation, which involves automatically translating text or speech from one language to another.

Gen AI: The Next Frontier

Now, gen AI represents a cutting-edge subset of AI that focuses on machines' ability to generate content often indistinguishable from that created by humans. One notable milestone in the evolution of gen AI is the development of models like OpenAI's GPT or Anthropic's Claude, which was first beta-tested in 2018. This model marked a transformative leap in the field by utilizing a massive neural network architecture trained on diverse data sets.

Gen AI transcends us into new territory of cognitive augmentation, which we characterize as "Human+." From aiding in brainstorming and refining drafts, to generating sample code and to offering real-time feedback, gen AI actively partners with us to facilitate a workplace where innovative outcomes become the norm. It empowers us to push the boundaries of what's possible.

The Human+ Workplace

In this Human+ phase of work, gen AI's contribution to lifelong learning is invaluable. Gen AI transforms our work dynamics, making us faster, smarter, and stronger. Imagine a world where gen AI acts as a constant teammate and assistant, freeing up human time for high-value activities like creative problem-solving and relationship-building. It can test new ideas and identify potential risks before human involvement, saving time and resources. As a mentor, gen AI provides personalized learning and real-time feedback, fostering continuous improvement and growth.

Gen AI provides advanced training and upskilling opportunities that seamlessly prepare the workforce for new challenges. With patience and objectivity, gen AI becomes an ideal learning facilitator and counselor, enabling personal growth and development at an individual's pace without fear of judgment. It democratizes access to knowledge and skills, ensuring that no one is left behind in the face of rapid technological advancements.

This partnership between humans and gen AI is not just about machines mimicking human abilities; it's about forging a relationship that amplifies our very essence, making us undeniably and beautifully human. It allows us to focus on what we do best—being creative, empathetic, and innovative—while leveraging the power of AI to enhance our capabilities. Together, we create a future where the possibilities are limitless, and human potential knows no bounds.

Here are just a few of the roles GenAI can have within your organization to truly take advantage of a Human+ workplace.

The Assistant

In 2007, Conversica founder Ben Brigham had a pioneering vision to utilize digital assistants for driving car sales. As an AI visionary, Brigham introduced AI-driven communication solutions that revolutionized customer engagement in the automotive retail industry, establishing new standards for interactions between businesses and their customers. While other AI companies focused on reducing costs through automated customer support, Brigham recognized the untapped potential of AI automation in generating revenue and making a significant impact on businesses' bottom lines.

Today, Conversica has evolved into a leading enterprise conversational AI automation platform, serving a wide range of industries and transforming the workforce models of revenue teams worldwide. The company has expanded its focus beyond sales teams to encompass all revenue teams, including marketing and customer success. By leveraging the latest advancements in generative and conversational AI technologies, Conversica empowers these teams to excel in three key areas:

1. *Generating and rekindling interest:* Conversica's AI enables teams to effectively engage new visitors, nurture cold leads, manage events, and re-engage former customers through sophisticated, dynamic conversations.

2. *Advancing lead engagement:* The platform facilitates the transformation of warm leads into high-value opportunities, efficiently qualifies inbound requests, and revitalizes dormant prospects.

3. *Strengthening customer relationships:* Conversica aids in managing customer health, renewing and expanding relationships through upsell opportunities, and gathering valuable feedback to ensure long-term success.

In an environment where revenue growth is hard, Conversica's revenue digital assistants (RDAs) don't give up when a human likely will. These RDAs excel at nurturing leads and guiding prospects through the sales pipeline, combining advanced AI capabilities with a human-like approach. Conversica's AI excels at understanding customer responses, automatically generating optimal replies and actions, and seamlessly integrating with leading customer relationship management and marketing automation platforms.

One notable example of Conversica's impact is with its customer, Leica Geosystems. Naming their automated messaging system 'Holly Hudson', the RDA takes a proactive approach to customer engagement by initiating personalized conversations with individuals who interact with the company's online materials. Holly identifies potential customers who express interest in learning more about Conversica's products or services. Holly maintains an ongoing dialogue with these prospects, providing them with relevant information and nurturing the relationship over time, even if they are not yet ready to make a purchase. This approach allows Leica to build a strong foundation for future sales opportunities and ensures that potential customers receive the support and information they need throughout their decision-making process. And the result?

Leica Geosystems achieved remarkable results: a 23x return on investment, a 33% lead engagement rate, a 300% improvement in converting net-new leads to marketing-qualified leads (MQLs), and a contribution of 12% to the total sales opportunity value. This success story demonstrates the power of Conversica's AI in identifying and addressing obstacles in the sales process, ultimately driving significant revenue growth for their clients.

Many organizations underestimate the cost of pursuing tedious tasks and fail to recognize the potential for AI automation to streamline these processes. For example, spending $3 million to generate $750,000 in revenue is not a sustainable approach. In the case of Conversica, however, AI automation can profitably reduce the $3 million expenditure to the cost of a single revenue digital assistant (RDA), making the $750,000 revenue target attainable. This translates to a revenue-based ROI of nearly 7x, a figure that both marketers and CFOs will applaud. Furthermore, when this is achieved without incurring additional workforce costs, the ROI has the potential to soar as high as 37x within a single year, showcasing the immense value that AI automation brings to organizations seeking to optimize their revenue generation efforts.

The Teammate

As we discussed at the start of the chapter, Asana's AI teammates were unveiled at the Work Innovation Summit in San Francisco and, at the time of the writing of this book, is still in beta testing. Asana AI teammates are adaptable collaborators that help teams maximize their impact and achieve goals faster. Infinitely customizable, AI teammates advise on priorities, power workflows, and even take action on work—all while adapting to the unique ways that individuals and teams work.

With humans in the loop every step of the way, people have full transparency and control over how AI teammates support work. Asana's introduction of AI teammates aligns seamlessly with cofounder Dustin Moskovitz's vision of creating a more human-centric work environment.

Powered by Asana's Work Graph data model, AI teammates offer unparalleled customization, allowing workflows to be tailored for every role, process, and industry. These AI teammates can even be used in a multiplayer state, enabling multiple people to collaborate with AI in real time on the same work. Asana AI teammates, can, for example:

- *Advise teams where to focus* by surfacing insights around potential risks to achieving goals
- *Produce work and workflows* at scale by taking on tasks, triaging requests, and assigning tasks
- *Adapt to how you work* by identifying where workflows are broken or could be improved based on organizational best practices and sharing relevant resources to inform work

In the creative production domain, for example, a leading outdoor advertising company is leveraging AI to transform its creative request process, allowing its team to focus on creative flow and design. AI teammates are tasked with triaging incoming requests, proactively gathering missing information, assigning work to specific people based on context, assisting with initial client research, and improving reporting quality through consistent data.

Similarly, a global cybersecurity company is utilizing AI teammates to streamline the execution of projects and campaigns, freeing up workers to focus on more strategic tasks. In this context, AI teammates are responsible for crafting on-brand, tailored marketing content, translating content using the company's internal library of brand terms, highlighting key information about incoming requests for faster prioritization, and enforcing naming conventions to improve cross-tool compatibility.

Moskovitz envisions a future where AI seamlessly integrates into every facet of work, not only boosting efficiency but also emphasizing and enhancing the fundamentally human aspects of our jobs. By empowering teams to collaborate effortlessly, AI paves the way for great accomplishments to become not just possible, but inevitable. Moskovitz believes that when we work together in harmony with AI, we unlock the potential to achieve remarkable things—things that contribute to a better life for people and the planet.

The Decision-Maker

In today's fast-paced business world, the volume of decisions made by managers across organizations worldwide is staggering. As businesses have grown increasingly complex, the need for rapid, data-informed decision-making has become paramount. Fortunately, AI has emerged as a powerful tool, equipping companies with insights that enable faster, smarter decisions and provide a competitive edge.

Marketing, once considered an art form, has been transformed by AI. Algorithms now analyze vast amounts of data, including consumer behavior, purchase histories, and social media activity, to predict which messages will resonate with specific customer segments. Coca-Cola, for example, leveraged AI to analyze 120,000 daily social media posts and adjust its marketing strategies in real time, resulting in more personalized engagement and improved sales.[4]

Moreover, according to Salesforce's "State of Marketing" report, 84 percent of customers prioritize being treated like individuals.[5] Platforms like Copy.ai, powered by advanced NLP models, offer comprehensive content solutions that enable businesses to create highly personalized content in minutes, a process that once took hours or days.[6] This agility allows companies like Salesforce, Survey Monkey, and HubSpot to fine-tune their marketing strategies and launch hyper-targeted campaigns tailored to individuals' unique experiences within specific time frames. AI makes the personal touch scalable, a crucial factor in winning and retaining customers.

Beyond marketing, AI is revolutionizing industries like pharmaceuticals. By analyzing vast amounts of medical journals, patents, and clinical trials, AI helps identify potential drug candidates, reducing the traditional twelve-year drug-discovery process by several years and saving billions of dollars. Accenture estimates that AI in drug discovery could bring cost savings of up to $26 billion for the health care industry.[7]

Supply chain optimization is another area where AI excels. Sophisticated algorithms can predict supply chain disruptions with remarkable accuracy, giving companies the foresight to navigate around global shipping delays or raw material shortages. A Gartner report highlighted that companies leveraging AI in supply chain planning saw a 7 to 13 percent improvement in revenue.[8]

AI-driven decision-making is the new frontier on which businesses will either seize a competitive advantage or get left behind. Whether tailoring marketing messages with unprecedented precision, revolutionizing R&D to bring lifesaving drugs to market faster, or optimizing supply chains to mitigate disruptions, AI is the modern-day pickax, capable of extracting valuable insights from the vast data landscape.

The Revolutionizer

The advent of gen AI is poised to fundamentally transform the way organizations operate, ushering in a new era of innovation and efficiency. While gen AI won't completely replace existing organizational structures, it will undoubtedly revolutionize business processes and redefine job roles. Rather than simply being incorporated into current positions, gen AI will drive the evolution and enhancement of job functions, leading to the emergence of new roles specifically designed to oversee, utilize, and optimize AI systems.

As gen AI becomes increasingly integrated into the workplace, we can anticipate a rise in cross-functional teams where human employees and AI work in tandem on projects. This symbiotic relationship will foster a culture of agility and innovation, enabling organizations to adapt quickly to changing market dynamics and stay ahead of the competition. With gen AI as a key player in the workforce, the emphasis will shift toward interpreting and acting upon the valuable insights it generates. In this context, several roles will take on heightened importance:

- Data analysts, AI strategists, and decision scientists will play a pivotal role in guiding data-driven decisions across all aspects of the business, leveraging the power of gen AI to uncover actionable insights and optimize performance.
- Specialists in AI governance and compliance will be crucial in ensuring that the use of AI aligns with ethical standards and regulatory requirements, maintaining public trust and mitigating potential risks.
- AI integration specialists will be responsible for seamlessly incorporating AI systems into existing workflows, ensuring that they align with organizational goals and deliver maximum value.

As organizations embrace gen AI, traditional hierarchical structures may give way to more fluid and adaptive models, with decision-making processes involving a blend of human expertise and AI-generated insights. This shift will necessitate a reimagining of learning and development initiatives, with dedicated roles focused on equipping employees with the skills needed to effectively collaborate with gen AI. Learning experience designers, AI trainers, and continuous learning managers will be instrumental in building a workforce that can harness the full potential of AI technologies.

Moreover, leadership roles will undergo a significant transformation, with a greater emphasis on collaborative leadership styles. Executives will be tasked with driving strategic initiatives that maximize the benefits of gen AI while fostering a culture of collaboration between human employees and AI systems. The integration of gen AI into the workplace presents a unique opportunity for organizations to unlock new levels of productivity, creativity, and innovation, powered by the symbiotic relationship between human ingenuity and AI.

The Democratizer

David Autor, the esteemed MIT labor economist hailed by the *Economist* as "the academic voice of the American worker," now recognizes modern AI as a "fundamentally different technology, opening the door to new possibilities."[9] In a thought-provoking article for the *New York Times*, Autor asserts that AI has the potential to "change the economics of high-stakes decision making so more people can take on some of the world that is now the province of the elite and expensive experts such as doctors, lawyers, software engineers and college professors. And if more people, including those without college degrees, can do more valuable work, they should be paid more, lifting more workers into the middle class."[10]

Autor's insights underscore a pivotal point: ChatGPT's ability to develop code empowers individuals and small teams to manage companies with unprecedented efficiency. Entrepreneurs who once relied on substantial capital investments can now launch ventures with minimal resources and reduced dependence on external support. AI models' capacity to generate code snippets and even entire programs based on natural language descriptions or prompts revolutionizes the coding process, accelerating development cycles and enabling developers to tackle coding challenges with unparalleled speed and precision.

This democratization of software development tears down barriers to entry for aspiring entrepreneurs, cultivating a more inclusive and accessible entrepreneurial landscape. As AI-powered code generation continues to evolve, it holds the potential to expedite the development of decentralized applications (dApps) and blockchain-based ventures, further fueling the engine of innovation in the tech industry. Brace yourself for a groundbreaking shift in the startup landscape: the imminent rise of a billion-dollar company, masterminded and helmed by a single visionary entrepreneur. This remarkable individual's singular vision, coupled with the transformative power of AI, will redefine what is possible for a company to achieve under the leadership of one person.

The integration of AI technology into software development processes represents a seismic shift in the future of entrepreneurship and technology-driven innovation. As AI capabilities expand and mature, they will undoubtedly reshape the contours of the business world, empowering a new generation of entrepreneurs to build groundbreaking products and services that transform industries and uplift communities. The era of AI-driven entrepreneurship is upon us, and its impact will be felt across every sector of the global economy.

A Holistic Approach to AI

As we approach the new gen AI–enabled world of work, we need to think about how to focus equally on the human in Human+. A holistic approach encompasses a comprehensive consideration of the societal, ethical, and human impacts of AI technologies. This approach emphasizes the importance of ethical and responsible AI development and deployment, prioritizing human rights, fairness, transparency, and accountability. Human-centered design principles are central, ensuring that AI systems are intuitive, accessible, and supportive of human capabilities and decision-making. Interdisciplinary collaboration is encouraged, bringing together diverse stakeholders to address complex challenges and ensure that AI technologies benefit society as a whole.

In tandem with prioritizing human-centered AI principles, it's imperative to recognize the profound influence gen AI could have on our collective health. By automating routine tasks and streamlining processes, gen AI will give us back valuable time that can be used to focus on our physical and mental well-being. This is particularly important given that modern health challenges such as increased stress, sedentary lifestyles, poor dietary habits, and environmental pollutants contribute to various chronic conditions like obesity, diabetes, and heart disease. As illness frequently results in absenteeism or necessitates employment-based health-care reliance, gen AI has the potential to enhance preventive health-care measures, personalize treatment plans, and improve health outcomes for individuals and communities.

One company at the forefront of this proactive approach to health-care is Biostate.ai. Cofounded by Ashwin Gopinath and David

Zhang, Biostate.ai leverages a systems biology approach to identify opportunities for repurposing previously unsuccessful drugs, enhancing their safety and efficacy for personalized treatment plans. As Gopinath, the chief technology officer of Biostate.ai, explains, "Biostate.ai is reducing the dependency on employment for health security and reshaping how individuals engage with work. This aligns with a broader vision where gen AI facilitates a more flexible and health-oriented approach to employment, significantly enhancing quality of life and work."

The innovative work being done by companies like Biostate.ai underscores the transformative potential of gen AI in the health care domain. By pioneering a future where health care is more accessible and tailored to individual needs, these advancements are paving the way for a more resilient and productive workforce. As we move forward into the era of Human+, it is clear that the symbiotic relationship between humans and AI will play a crucial role in shaping the future of work and our overall well-being.

A Shift in the Labor Force

Throughout history, various technological advancements have led to shifts in the labor market, causing certain job roles to become obsolete, while creating opportunities for new ones to emerge. One notable example is the Industrial Revolution, which unfolded over the eighteenth and nineteenth centuries and brought about significant changes in manufacturing and production processes. Innovations such as the steam engine, mechanized looms, and assembly lines revolutionized industries like textiles, agriculture, and transportation. While these advancements undoubtedly increased productivity and efficiency, they also displaced many traditional craft-based jobs, such as skilled artisans and handloom weavers who saw their livelihoods threatened by the rise of automated machinery in textile mills.

As AI continues to advance, its impact on the labor market is becoming increasingly apparent. According to a 2023 Goldman Sachs report, gen AI has the potential to eliminate three hundred million jobs worldwide, while also creating new opportunities in fields such as AI, machine learning, and ethical AI.[11] Another recent study by Goldman Sachs found that 60 percent of today's workers are employed in occupations that didn't exist in 1940, implying that more than 85 percent of employment growth over the last eighty years is explained by the technology-driven creation of new positions.[12]

As leaders navigate this shifting landscape, it is crucial to approach the adoption of AI not as a replacement for human talent but rather as an opportunity for growth and collaboration. By setting a holistic approach and guidelines that prioritize transparency, fairness, and accountability, leaders can champion the ethical deployment of AI technologies while supporting their workforce through the transition. This may involve investing in reskilling and upskilling initiatives, fostering cross-functional collaboration between human workers and AI systems, and actively engaging with employees to address concerns and identify opportunities for growth.

As we move forward into the era of Human+, it is clear that the success of our organizations will depend on our ability to foster a culture of collaboration between humans and AI. By embracing a human-centered approach to AI adoption, we can harness the unique strengths of both human talent and AI, driving innovation, growth, and positive change. As leaders, it is our responsibility to navigate this transformative journey with empathy, transparency, and an unwavering commitment to the well-being of our employees and the ethical development of AI technologies.

CONVERSATION STARTERS

1. In what ways do you foresee gen AI transforming traditional workflows and processes within your organization? What strategies can be implemented to effectively manage the transition?
2. What measures should your organization take to address potential job displacement or reskilling needs resulting from the integration of gen AI?

3. As gen AI becomes increasingly sophisticated, how can you ensure transparency and accountability in AI-driven decision-making processes, to both internal teams and external stakeholders?
4. What role do you see gen AI playing in fostering innovation and driving competitive advantage for your company?

CHAPTER 4

The Transformative Power of Web3 Technologies

Nicholas Inlove had always been a trailblazer in the tech sector, and with a keen eye for emerging trends, he decided to test out Juicebox DAO—a platform similar to a decentralized version of Kickstarter.[1] Described as a "blockchain-empowered Co-op," Juicebox DAO governs its own financial, organizational, and operational aspects.[2] But what piqued Inlove's interest most was the unique term "contributors," an alternative expression used in the industry to replace the word "employee."

In the world of Juicebox DAO, traditional job titles like CEO, manager, and intern were rendered obsolete. Instead, workers were deemed contributors, with the power and agency to contribute meaningfully to the collective vision. The emphasis was on the value a person brought to the community rather than their place in a hierarchical pecking order. Compelled by this egalitarian framework, Inlove joined the DAO and soon found himself fully immersed in projects that aligned seamlessly with his expertise and interests.

As he deepened his engagement, Inlove realized that Juicebox DAO was more than an alternative to traditional employment. It was an entire ecosystem built on communal values. Decision-making was collaborative, facilitated through the use of tokens as a means of democratized choice. Transparency was hardwired into the community's operational culture. And the flexibility to contribute wherever and whenever offered Inlove a liberating departure from the limitations of conventional work environments. Juicebox DAO provided Inlove with the dynamic, transparent, and decentralized community he had long sought—a place where he could actively nurture his career passions.

The same principles that made Juicebox DAO so appealing to Inlove are foundational elements of this next phase of the internet. Simply known as web3, this evolution is an extraordinary shift from centralized to decentralized digital infrastructures. And just as Inlove found his professional tribe in a decentralized community, web3 promises to radically reshape how we all engage, transact, and work in digital environments.

What Is Web3?

Simply put, web3 is the next iteration of the internet. It represents a significant departure from the conventional, centralized frameworks we are accustomed to by embracing a decentralized, user-centric approach. In contrast to web 2.0, where corporations possess user data, web3 leverages blockchain technology, decentralized applications (dApps), and smart contracts to empower individual contributors. This transformative shift in the digital landscape is forecasted to propel the web3 market to a projected value of approximately $53 billion by 2030.[3]

In the early nineties, web 1.0 served as a sort of digital billboard where businesses and individuals could publish static pages—essentially, digital brochures—that viewers could read but not necessarily interact with. While revolutionary for its time, the first iteration of the web was essentially a one-way street. It was called the "information superhighway" for a reason, great for surfing through information but lacking off-ramps for user engagement.

Fast-forward to the early 2000s, and the internet morphed into something much more interactive—web 2.0. This was the era of social media, blogs, and online marketplaces. User-generated content became king. Suddenly, the internet was not just a library but a bustling town square. Yet, it wasn't without its problems. As companies like Google, Facebook, and Amazon grew, they became like giant malls, exerting immense control over user data and transactions. These centralized entities owned the user-generated content and held the keys to the user experience, which, as you can imagine, poses serious questions about data ownership, privacy, and economic equity.

That brings us to the emergence of web3. If web 1.0 was a library and web 2.0 a bustling town square, then web3 can be compared to a decentralized city, complete with its infrastructure, economy, and governance (see figure 4-1). Most notably, it lacks a centralized authority figure, like a mayor. Built on blockchain technology, which is essentially a secure, transparent, and immutable digital ledger, web3 operates on principles of decentralization. The aim of this technological movement is to return control and data ownership to individual users, all while providing a potent platform for collaboration, business, and innovation.

Each version of the internet brings new ways of working. Web 1.0 allowed us to work from anywhere. Web 2.0 empowered entrepreneurs via social media (even if the returns often rebounded to the platforms). And now web3 is creating ways for people who don't know one another to collaborate on open-source projects and distribute the collective returns equitably.

FIGURE 4-1

The evolution of the web

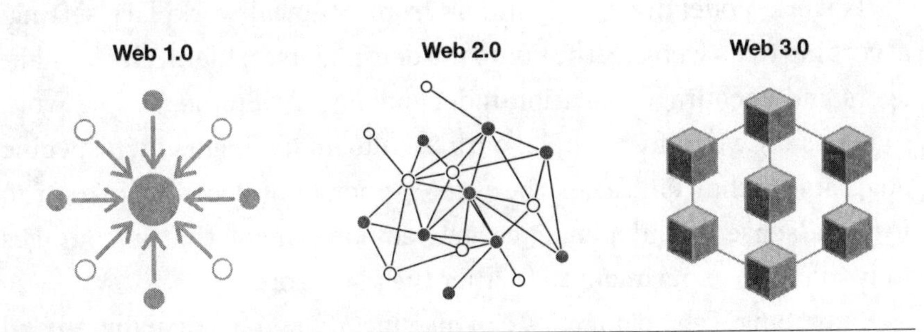

Why Is Web3 Revolutionary for Work?

Why do we need web3 for more efficient and meaningful work? For one, it eliminates a lot of the red tape commonly associated with centralized systems. With blockchain's secure, transparent ledger technology, costly middlemen can be removed from many types of transactions—whether that's buying a home, securing a supply chain, or validating someone's professional credentials.

Imagine a supply chain scenario in which components are sourced from multiple countries. Typically, this would involve a nightmarish tangle of paperwork, compliance checks, and financial transactions. With web3 and smart contracts—self-executing contracts with the terms directly written into code—all parties can automate and verify each step of the process in real time, dramatically reducing both time and costs.

Beyond efficiency, web3 also brings a new level of meaning to work by enabling a genuine gig economy where individuals can sell their services directly to consumers without an overseeing platform taking a large cut. Think of it as Etsy or Upwork, but without centralized control, where artisans and freelancers have full ownership and control over their economic interactions directly with consumers.

Consider the company Status, an open-source mobile app that serves as a web3 browser and messenger on the Ethereum blockchain, cofounded by Jarrad Hope and Carl Bennetts. The app allows users to interact with dApps and Ethereum smart contracts directly from their mobile devices.

Its work model drastically differs from a typical workplace setting: according to its website, the company doesn't have a hierarchy or managers and encourages location independence. At Status, people work in teams of two to five people, with each team having its own specific goal. Status thus embraces the web3 principle of decentralization to foster a sense of independence and empowerment that encourages individuals to experiment to further the platform.

Embracing web3 principles can also mean having a distributed and remote workforce. Because Status encourages location independence and implements a strictly remote work model, a seemingly infinite number of members from different parts of the world can collaborate on projects using digital communication tools and blockchain-based collaboration platforms. This approach fosters inclusivity and diversity as it taps into the global talent pool.

The success of Status's remote work environment showcases the company's readiness to embrace risk and venture into uncharted territory, key factors in the effectiveness of its web3 work model. While few traditional web 2.0 companies have dared to transition from in-person offices to a fully digital, decentralized setup, this hesitation is precisely what limits their progress.

Web3 fundamentally reimagines what work can represent—a more democratic, equitable, and user-centric ecosystem—and as we increasingly migrate our work and lives onto digital platforms, web3 provides the tools and ethos to make the world of work more efficient, transparent, and equitable. For business leaders, now is the time to understand and leverage this groundbreaking technology by embracing its capacity to reshape the very essence of how we work and interact.

Applications of Web3

Web3 will massively shift how we interact with digital platforms by changing our economic relationship with them. For instance, imagine participating in a decentralized finance (DeFi) platform through your company, where instead of your retirement savings sitting in an often predetermined set of investment options, they would be actively engaged in revenue-generating activities, offering you potentially higher returns. Gartner research predicts that by 2029, business value added by blockchain will soar to just over $176 billion and then exceed $3.1 trillion by 2032.[4] That's a staggering number that signals the enormity of the economic shift that web3 could facilitate.

What does this mean for the digital-first leader? For starters, embracing web3 technologies can significantly enhance your organizational efficiency, cut costs, and amplify innovation. No longer will you have to negotiate with multiple vendors for services that can be securely and efficiently executed on a blockchain. Your HR department could potentially oversee a global workforce as easily as a local one, thanks to decentralized identity and credential verification systems.

Your challenge, then, will be to understand how to harness this distributed value to benefit your organization and workforce. And while this may require a shift in your leadership approach, the rewards—increased innovation, efficiency, and engagement—are more than worth the effort. Web3 represents a groundbreaking transformation of the existing internet that aims to create a digital landscape that is more democratic, secure, and user-friendly. Business leaders who choose to view this transformation through a lens of opportunity rather than as a threat are the ones who will successfully navigate this change to democratize the future of work.

How Web3 Energizes the Daily Grind

Let's take a step back for a moment and view employment from a thirty-thousand-foot view. Work, as it stands today, often feels like a never-ending treadmill. We're constantly chasing higher productivity and better efficiency, and paradoxically, we're ending up less satisfied and more burned out. Wickham Skinner, the late Harvard Business School professor, hit the nail on the head when he said that the more we obsess over productivity, the more elusive it becomes. His insights reflect a significant crisis in today's workforce—a crisis of fulfillment.[5]

Employee disengagement is a well-documented problem. A staggering 77 percent of Americans have experienced burnout, leading to physical fatigue in 44 percent and cognitive exhaustion in 32 percent.[6] This "always on" culture is grinding down individuals, which in turn grinds down entire organizations. We're talking about reduced productivity, higher turnover rates, and a toxic organizational culture that no number of casual Fridays can fix.

Web3 technology promises to make work intrinsically motivating again. No more drudgery of hierarchical decision-making, or opaque policies that no one understands. In this world, instead of your daily tasks being handed down from on high, you choose challenges and projects that align with your skills and aspirations. This democratization of work goes beyond the potential of increasing employee engagement by redefining what we consider to be productive work altogether.

For example, web3 can help create decentralized teams that collaborate on projects of genuine interest and societal value, and not just tasks that fit into a quarterly financial report. Picture engineers, marketers, and customer service reps from across the globe, all collaborating on a project they're passionate about, their roles and rewards clearly outlined within immutable smart contracts. When we give people the freedom to self-organize and choose their contributions, we are simultaneously making work more enjoyable and more effective.

So, let's step off the treadmill and take a moment to consider the potential of web3. What if our chase for productivity has been misguided? What if, instead of asking how we can do more, we start asking how we can do better—and find more fulfillment in the process? It's time to question the traditional career path set before us since grade school. The emerging web3 era provides a captivating alternative, suggesting that maybe, just maybe, life—and work—can be much more meaningful than we ever imagined.

The Promise of Web3 in Revolutionizing Employee Experience

HR professionals have seen their fair share of trends aimed at enhancing the employee experience, yet despite their best efforts to design workplaces that cater to the multifaceted needs of their workforce—through flexible hours, workplace wellness programs, or employee engagement initiatives—they find themselves up against the limitations of a system built for a different era. The traditional corporate structure can be rigid and hierarchical, causing inevitable friction and dissatisfaction among employees. This is why the transformative promise of web3 will endure beyond the hype.

But the real transformation happens when web3 is applied to the employee experience directly. Rather than a top-down management structure where decisions are made at the C-suite level, decentralized organizations empower all members to have a stake in decision-making, from work policies to project prioritization. This is participative management on a whole new level, enabled and safeguarded by blockchain technology. This new approach has the power to boost employee engagement by finally transforming employees into vested stakeholders with a meaningful say in their work environment and assignments.

Smart contracts can program compensation, one of HR's perennial roles, to administer real-time, performance-based rewards, entirely removing the need for cumbersome annual reviews. In a world where personalized experiences are becoming the norm, the one-size-fits-all model of benefits and bonuses can finally make way for an individualized work experience.

Then there's the issue of career development and skill-building, a critical component of employee satisfaction and long-term retention. Decentralized, skill-specific marketplaces can facilitate microlearning and growth. This personalizes the upskilling journey for each individual and creates a culture of continuous improvement and meritocracy.

Of course, all these advancements come with challenges, but the promise of web3 could extend far beyond the tech department and has the potential to solve some of the most persistent challenges for the modern workforce. From a more equitable power distribution to personalized compensation and continuous growth, the possibilities are transformative. As stewards of organizational culture and employee well-being, we have a unique opportunity to be at the forefront of this change, steering our organizations toward a future where work is more than a contract—it's a collaborative experience for everyone involved.

The Transformative Power of Blockchain

To understand the incredible promise of blockchain technology, imagine a highly secure, transparent digital ledger that no single entity controls but everyone can trust. This technology enhances transparency, drastically lowers the costs of trust and coordination, and allows for decentralized consensus (see figure 4-2). Ethereum, for

FIGURE 4-2

Blockchain transaction process

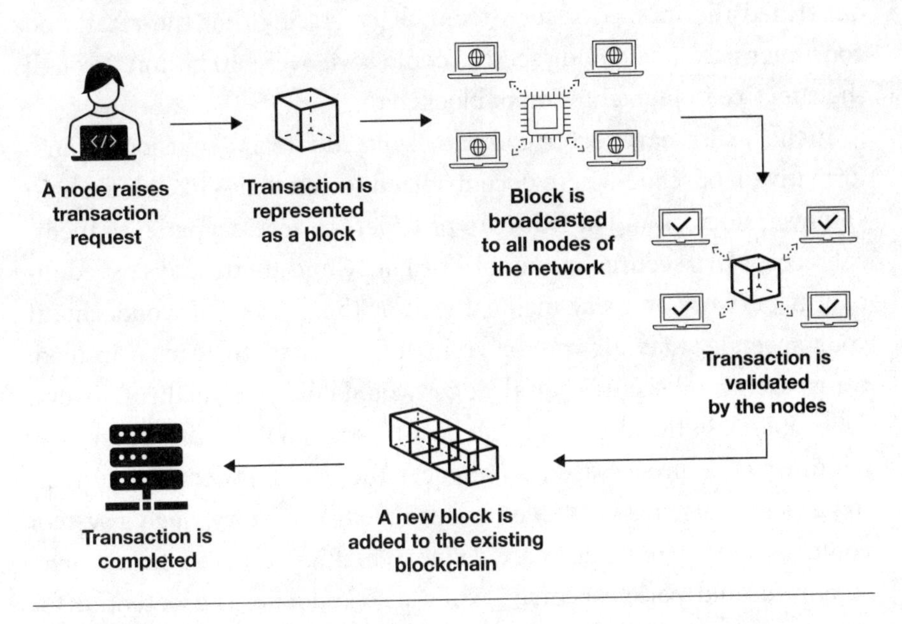

example, the brainchild of Vitalik Buterin, is the most popular and widely used blockchain, which establishes a peer-to-peer network that securely executes and verifies application code, called smart contracts (addressed more thoroughly in chapter 7). Ethereum offers an extremely flexible platform on which to build decentralized applications using the native Solidity scripting language and Ethereum Virtual Machine.[7] Decentralized application developers who deploy smart contracts on Ethereum benefit from the rich ecosystem of developer tooling and established best practices that have come with the maturity of the protocol.

The following examples show how companies are leveraging blockchain technology. Executives of well-known luxury watch companies like Rolex, Omega, and Patek Philippe have been battling a problem with counterfeit products for years that have plagued their market share and harmed their brand reputation.[8] In these cases, blockchain technology makes it possible to trace the provenance of each watch component to its origin, securing the entire supply chain against fraud. This level of traceability and transparency can virtually eliminate counterfeits. A 2021 report by Capgemini Research Institute quantified the impact of such technology, stating that the retail and consumer packaged goods sectors could save over $30 billion through the effective implementation of blockchain.

In the health-care sector, companies like Medrec:M envision a transformative landscape where decentralized technologies like blockchain empower doctors and health-care providers to access a patient's medical records in a secure, immutable format.[9] Gone are the days of redundant paperwork at every medical visit or the anxiety of wondering if your specialist has all your current information. Instead, a unified, transparent, and secure digital ledger would hold your medical history, accessible only by authorized medical professionals. This streamlines administrative processes and enhances the quality of care by providing a comprehensive view of a patient's medical history. Such a system could revolutionize health-care delivery, making it more efficient, personalized, and patient-centric, while also reducing the potential for errors and improving overall patient outcomes.

It's not hard, then, to see the profound implications of blockchain technology for employment. Leaders can leverage blockchain to streamline and enhance various aspects of workforce management, from recruiting and onboarding to payroll processing and performance reviews. Research from Deloitte's 2024 tech trend report shows that 81 percent of executives believe that blockchain technology is broadly scalable and will achieve mainstream adoption.[10] Blockchain technology offers HR leaders an unprecedented opportunity to rebuild and enhance their workforce management systems, making them more adaptable, secure, and aligned with the decentralized web3 world of work.

The Evolving Role of Leadership in the Age of Web3

With the advent of web3, companies will find that their newest team could actually be a DAO, composed of thousands of contributors globally. They vote on company decisions, propose changes in real time, and can execute complex projects without waiting for quarterly reviews or managerial green lights. The organization has transformed into a dynamic ecosystem where the traditional pecking order is flipped on its head. This new workforce could be more engaged because the workers are stakeholders, as opposed to salaried workers, with a tangible share in the collective outcome.

In this new landscape, the role of business leaders evolves as well. No longer the ultimate authority, you become the facilitator of a decentralized network, ensuring that governance tools are in place and that every voice can be heard. You are not threatened by this loss of unilateral control because you see the incredible uptick in innovation and problem-solving capacity.

As you venture into this new and unknown territory, how well you adapt and grow will shape your impact in the future. This is a new ideal where decentralization and collective intelligence take center stage, inadvertently tearing down the walls of traditional corporate hierarchy to erect a more inclusive, equitable, and efficient form of employment.

How Web3 Leadership Will Work

A decentralized organizational structure is characterized by the delegation of decision-making authority from senior management to lower levels within the organization. In this setup, managers overseeing cost centers, profit centers, or investment centers have the autonomy to make decisions that directly impact their specific areas of responsibility. Some decisions may even trickle down to individual employees, although these are typically limited to matters related to customer service, such as the discretion to offer free shipping to a customer. A decentralized organizational structure proves effective in the following scenarios:

Highly individualized customer service. This is particularly suitable when a high degree of personalized customer service is required, especially at the point of direct interaction with customers. It allows employees to adapt to unique customer needs and preferences.

Numerous store locations. In cases where an organization operates multiple store locations, a centralized approach becomes impractical. Decentralization enables each location to make localized decisions, as senior management cannot feasibly oversee or manage all locations individually.

Intense competitive environment. Industries marked by fierce competition demand swift and dynamic decision-making. A decentralized structure empowers various units within the organization to make prompt decisions in response to competitive actions, helping the company maintain agility in a rapidly changing landscape.

Constant business model evolution. When an organization's business model is subject to frequent innovations and changes, centralized control may hinder adaptability. Decentralization allows different parts of the organization to respond independently to evolving circumstances, ensuring that the business remains flexible and responsive to market shifts.

Highly personalized marketing. Decentralization is increasingly vital in marketing strategies that aim for a high level of personalization. It enables individual teams or units to tailor marketing campaigns to specific customer segments, leveraging real-time data and analytics. This approach facilitates more targeted and effective marketing initiatives, adapting to consumer behavior and preferences on a granular level, which would be challenging to manage centrally.

As we transition from the fundamentals of decentralized organizations, our focus shifts toward the pioneering management frameworks of web3, which are reshaping the future of work. In the following section, we explore the application of these groundbreaking technologies in crafting management structures that empower every employee. By redefining governance and value creation, web3 technologies open the path toward a new era of organizational operation and employee empowerment.

A New Management Structure

In the fall of 2021, Austin Cain and Graham Novak set out with a bold vision to crowdsource funds to acquire a rare copy of the US Constitution via a DAO named ConstitutionDAO.[11] Launching their initiative on Discord, a platform favored by web3 aficionados, they quickly galvanized a community of over eight thousand members, collectively raising $47 million in Ethereum within a week. Despite a modest median contribution of $217, the effort underscored the potent financial leverage achievable through collective action. The governance token, aptly named $PEOPLE, and the requirement for a consensus mechanism for transaction approvals—mirroring the constitutional ratification process—further emphasized the project's democratic ethos.

Despite its innovative approach and substantial fundraising, ConstitutionDAO was outbid by hedge fund manager Kenneth Griffin at the Sotheby's auction, due in part to the auction house's restrictions on DAO participation. Nevertheless, the project's impact went beyond its initial goal, redefining $PEOPLE's value from fractional ownership to a broader focus on governance. ConstitutionDAO not only highlighted the potential of DeFi and blockchain for cultural preservation but also marked a shift toward community-driven philanthropy. By fostering a widespread discussion on the cultural significance of historical artifacts and demonstrating the possibilities for decentralized funding in cultural and philanthropic endeavors, ConstitutionDAO exemplified the transformative potential of blockchain technology in fostering community involvement and reimagining collective support for cultural heritage.

How does this relate to integrating new management structures? In a world of increasing complexity and rapid change, centralized decision-making often becomes a bottleneck. According to a 2023 McKinsey report, companies that prioritized decision-making agility were more than twice as likely to have a top-quartile earnings-before-interest-and-taxes performance.[12] Decentralization has the potential to greatly enhance organizational agility.

Though we devote an entire chapter to DAOs later in the book, understanding them as a practical application within the context of web3 principles helps to frame how drastically the relationship to work is shifting. The implications range from more engaged workers and quicker decision-making to broader global collaboration. As a leader, you have a chance to leverage this evolutionary step to magnify your influence and effectiveness.

Decentralized Corporate Governance

The MolochDAO serves as an active case study that not only breaks the norm but shatters the ceiling of traditional management oversight. Operating within the Ethereum ecosystem, MolochDAO has leveraged the power of collective decision-making to such a degree that it's been able to funnel millions of dollars toward various Ethereum projects—all without a conventional hierarchical structure, board of directors, or CEO to guide its actions. What makes this possible? It's a governance model that puts the power back in the hands of the stakeholders themselves. Each member has a direct say in the allocation of resources, which are funneled through the DAO's smart contracts to worthy initiatives. By cutting through the bureaucracy typically involved in such decisions, MolochDAO has significantly accelerated the process of innovation within the Ethereum community.

The success of MolochDAO highlights the potential advantages of decentralized governance: agility, direct stakeholder involvement, and a fast, transparent decision-making process. The spotlight is on technologies that can not only deliver economic efficiencies but also reimagine the very essence of organizational structure and corporate governance. If a "company" like MolochDAO, devoid of traditional management and oversight, can effectively allocate millions of dollars in resources through collective decision-making, what other aspects of business can be reengineered for efficiency, transparency, and broad-based stakeholder engagement?

The impact of this transformative approach doesn't stop at processes and protocols. It has a profound effect on one of the most valuable assets of any organization: its people.

Augmenting Leadership in the Web3 Era

If you're an executive who's skeptical about web3, you might be concerned that all this talk of decentralization, AI, and peer-to-peer networks could put you out of a job. After all, if decisions are made at the top and people are choosing their own work, what's left for you? Let's dispel this concern right now: web3 technologies are not a threat but a powerful ally to equip you with new tools and frameworks to lead more effectively, make more informed decisions, and foster a culture of accountability and innovation within your organization.

Consider transparency, one of the ten operating principles of decentralization and arguably one of the most sought-after traits in modern corporate leadership. Imagine the level of trust and transparency you could establish, not just within your team but also with shareholders and customers. No more speculations or accusations; everything is transparent and verifiable. According to a Deloitte report, executives who embrace blockchain technologies for enhancing transparency and traceability stand to significantly streamline their operations, boosting both transparency and corporate reputation.[13]

What about decision-making? Think about self-executing contracts guiding corporate strategies or automated consensus algorithms aiding in decision-making processes. Imagine launching a new product and having real-time, reliable data about every aspect of production, from raw material sourcing to customer satisfaction, all verified and available at your fingertips. You're no longer relying solely on middle managers to inform you; instead, you're making data-driven decisions based on comprehensive, immutable data.

And let's not forget about team-building and accountability. One of the biggest struggles for any leader is ensuring that everyone on the team is accountable and aligned with the company's goals. In a decentralized organization powered by web3 technologies, every team member can have a clearly defined, transparent role and set of responsibilities, recorded and incentivized through smart contracts. HR systems where performance reviews, rewards, and even promotions are executed transparently through decentralized mechanisms are now viable. Employees are accountable to their immediate supervisors as well as the entire organization.

Web3 isn't about replacing leaders. It's about augmenting leadership. With these technologies, you're not becoming obsolete—you're becoming optimized. It's an opportunity to lead in a way that's more open, more democratic, and ultimately more effective.

The Unfolding Uncertainties of Web3

In our discussions with global leaders surrounding web3, we are often met, understandably, with skepticism. Some view it as a distant utopia that has yet to be proven on a large scale, while others dismiss it as the hopeful vision of those disillusioned with the current system. When Bill Gates was first invited to the *David Letterman Show* to share the concept of the internet, a skeptical audience couldn't look past its application as it relates to the radio. Similarly, these technologies are so new and transformative that mainstream consulting models have yet to fully grasp their true potential and application in a meaningful and transformative way.

However, the architects and advocates of web3 truly relate to the credible concerns leaders raise. How do we protect our identity? How do we stay rooted in productive work? How do we monitor and manage the exchange of digital currency and goods?

One of the foremost challenges lies in the regulatory and legal landscape. As web3 technologies begin to redefine what work looks like, existing legal frameworks struggle to keep pace. Questions around employee rights, contract enforcement, and liability in a decentralized ecosystem are yet to be fully addressed. This uncertainty poses a significant challenge for HR professionals and business leaders, who must navigate these uncharted waters without clear guidance or precedent.

Even more pressing is the issue of digital divide and access inequality. While web3 has the potential to democratize work opportunities globally, it also risks exacerbating existing disparities in digital access and literacy. Ensuring that the benefits of web3 technologies are accessible to all, regardless of geographic location or socioeconomic status, remains a pressing concern. This includes providing the necessary training and resources to equip a diverse workforce with the skills needed to thrive in a web3-dominated future.

Another uncertainty revolves around the impact of automation and AI on employment. While these technologies can enhance efficiency and productivity, they also raise concerns about job displacement and the need for significant workforce reskilling. The challenge for leaders will be to strike a balance between leveraging automation for organizational benefit while also investing in the development and transition of their human workforce to new roles and opportunities.

Finally, the psychological and cultural adjustments to a web3 work environment cannot be understated. The shift toward decentralized decision-making, remote work, and digital interactions requires a fundamental change in organizational culture and individual mindset. Building trust, fostering collaboration, and maintaining a sense of community in a digital-first workspace are critical challenges that leaders need to address to ensure the successful integration of web3 technologies into the workplace. As we continue to explore and adapt to these new models, leaders need to remain agile, open-minded, and proactive to guarantee a smooth transition for their organizations and their workforce into the web3 era.

Preparing for the Inevitable Web3 Transition

Inertia is the greatest enemy in times of technological transformation. Think back to the transition from brick-and-mortar retail to e-commerce. Many traditional retailers hesitated to adapt, holding on to the narrative that customers would always want the in-person shopping experience. Well, where are they now?

A study by Capgemini in partnership with the *MIT Sloan Management Review* shows that digital leaders—companies that managed to adapt to new technologies—outperform their competitors in every metric, from market share to revenue and profitability.[14] Early adopters are already seeing the dividends, and if history is any indication, they will be the market leaders of tomorrow. So, the question you, as a business leader, need to ask is not if you should adapt, but how willing and able you are to stay ahead.

Leaders who adapt and adopt the web3 philosophy will find that it perfectly dovetails with the key tenets of effective leadership—having a vision, ensuring transparency, and solving problems collectively. A 2022 study by *MIT Sloan Management Review* asserts that companies that adapt to technological change, in particular digital currencies and blockchain, can expect an uptick in financial performance and customer satisfaction.[15] Web3 is a philosophical shift toward a more open, transparent, and equitable way of doing business.

As we conclude our exploration of how web3 technologies are reshaping the landscape of work, it's clear that we're on the brink of a revolutionary transformation. Across various sectors, forward-thinking organizations are already leveraging the decentralized, transparent, and secure nature of web3 to redefine their operational models and employee engagement strategies. Early adopters view web3 as a definitive answer to their present employment challenges and are actively crafting a more promising future of work.

Imagine a team of contributors, as invigorated as Nicholas Inlove, who get to work when they want from wherever they are in the world and have enough say in the direction of the project to make meaningful contributions in a way that brings fulfillment back into their chosen career. This level of engagement has the power to supercharge productivity. Companies that integrate these technologies into their workflows benefit from increased efficiency, stronger security measures, and a more engaged workforce. By empowering contributors with more autonomy and a voice in decision-making processes, organizations can achieve unparalleled levels of productivity and satisfaction. The time to embrace web3 is not tomorrow; it's today. So let's make work better for everyone involved.

CONVERSATION STARTERS

1. In a world where the concept of "employee" evolves into "contributor," and decision-making is democratized through token-based governance, what new leadership qualities and skills will be most important for guiding such organizations?
2. How can you as a leader adapt to a role that emphasizes facilitation and empowerment over direct control, and what strategies can you employ to foster a culture of innovation, accountability, and collective achievement?
3. What ethical frameworks and governance structures do you need to develop to address the challenges posed by this transition, ensuring that the benefits of web3 are equitably distributed and aligned with broader societal values?
4. As your organization moves toward a more decentralized, flexible, and transparent work environment, what are the implications for the future of employment, skill development, and career paths?
5. How can individuals and organizations prepare for these changes, and what opportunities does web3 present for creating more meaningful, engaging, and fulfilling work experiences?

CHAPTER 5

The Immersive Internet: The Metaverse, Digital Twins, and Virtual Reality

Neha Singh, CEO and founder of Obsess, an immersive shopping platform launched in 2017, has always been a problem-solver. Growing up in Abu Dhabi, she was educated in an Indian school where computer science was one of three tracks students could study. "I always loved breaking down problems logically. With coding and computer science, I felt I had the power to build anything and solve any problem," she said.[1]

After graduating from the University of Texas, Austin, and completing a master's in computer science at MIT, Singh worked as a software engineer at Google and then led engineering and product development at the e-commerce startup AHAlife. She then served as *Vogue*'s first head of product for digital business, overseeing content, ads, and distribution platforms. It was at *Vogue* that Singh recognized that the whole world, even fashion, now runs on software. Having a front-row seat for luxury brand investments in their physical customer experiences like flagship stores and fashion shows, Singh saw an enormous void in how the brand experience and products appeared online.

"I kept thinking, why are ecommerce experiences so limited? And how the ecommerce experience hasn't evolved in twenty-five years when it was first created to sell books. Shopping for products online was like scrolling a database, where there is no feeling and no emotional connection to the brand. This is counterintuitive to all the work, materials, and creativity of the designers and makers who put their passion into it," said Singh.[2]

Singh's aha moment came with 3D technology, which became the genesis for Obsess. She figured out a different way to tell the story for brands that needed more beyond the limited options of font and color changes currently offered on e-commerce platform templates. In doing so, she aimed to provide visually and emotionally rich immersive shopping experiences, making products and storefronts so real-looking that you feel like you are actually there. It took eighteen months, $14 million in investment, a patent filing, and a lot of hard work to get a digital store to feel photorealistic.

The first big client was Tommy Hilfiger, whom the Obsess team met at ShopTalk in 2019, a conference where retail, innovation, and tech intersected. They created a virtual pop-up using the Obsess platform and subsequently five other immersive experiences based on the success of the first one. This established the first great partnership in the immersive shopping space. With the Covid-19 pandemic, demand surged for the Obsess platform, and having an online presence that went beyond a database became a necessity for brands.

Then, Apple came along with the Vision Pro—the first spatial computer. Obsess launched four fully immersive experiences in the shopping category with the launch of the device in February 2024, featuring classic American fashion brand J.Crew, wellness brand Alo Yoga, the trending beauty brand e.l.f. Cosmetics, and luxury retailer Mytheresa. These experiences showed what the future of digital commerce would look like, and how the digital customer experience across any vertical would transform with 3D.

. . .

The "immersive internet" is the next generation of the internet that leverages 3D and other interactive technologies such as augmented reality (AR), virtual reality (VR), and mixed reality (MR) to create highly engaging, lifelike, seamless, and immersive digital experiences. Focusing on cutting-edge concepts such as the metaverse, digital twins, AR, and VR, we explore how these disrupting technologies are reshaping the way companies operate, collaborate, and engage with their stakeholders.

Corporate leaders must grasp the transformative potential of these tools to stay ahead in today's competitive market. The metaverse, for instance, represents a paradigm shift in digital interaction, offering immersive, interconnected virtual spaces where individuals can work, socialize, and create. Digital twins provide businesses with real-time simulations of physical assets, enabling predictive maintenance, performance optimization, and enhanced decision-making. AR and VR technologies offer immersive experiences for training, product demonstrations, and customer engagement, fostering deeper connections and unlocking new avenues for innovation.

Rich virtual office spaces are poised to revolutionize the concept of traditional workplaces, offering dynamic environments where teams can collaborate seamlessly across geographical boundaries. Corporate leaders need to understand the potential of these technologies to drive efficiency, agility, and innovation within their organizations. Embracing the metaverse, digital twins, AR, VR, and virtual office spaces can position companies at the forefront of industry disruption, unlocking new possibilities for growth and success.

The Metaverse

For all its futuristic hype, the metaverse is already here and set to reshape the way we work, interact, and innovate. As Matthew Ball describes in his book, *The Metaverse: How It Will Revolutionize Everything*, the metaverse is a "massively scaled and interoperable network of real-time rendered 3D virtual worlds which can be experienced synchronously and persistently by an effectively unlimited number of users with an individual sense of presence, and with continuity of data, such as identity, history, entitlements, objects, communications, and payments."[3]

The metaverse transcends the limitations of physical reality, offering immersive and interactive experiences where individuals can work, play, socialize, and create. In the context of work, the metaverse is shaping the future of collaboration by providing innovative solutions for remote teams, enabling seamless communication, collaboration, and productivity regardless of geographical barriers. Through virtual meetings, conferences, and workspaces, the metaverse facilitates a sense of presence and connection, enhancing teamwork and creativity.

The adoption of metaverse technologies isn't going to be a "big bang" moment where we all suddenly find ourselves strapped in a headset reporting to hologram bosses. Rather, it will be a gradual, incremental process that will introduce us to a myriad of digital experiences transforming how we live, work, and interact. It's much how smartphones went from novelties to an indispensable tool for daily life. That didn't happen overnight for most of us, as it did for early adopters, but the shift was revolutionary, nonetheless.

Product developers worldwide are already working to tackle this by making their headsets lighter and more comfortable. On the software side, there's a move to make VR experiences more social and interactive. Apple's Vision Pro aims to make VR meetings as natural as face-to-face interactions, thereby reducing the feeling of isolation. The right solution to a specific business problem remains key to a successful AR or VR implementation. While business metaverse solutions that are fully immersive seem best for collaboration, providing customized environments, MR or AR solutions seem more applicable when individuals want to increase their individual productivity at their workstation. Generally, the initial accessibility and acceptance of "metaverse light" experiences that are less immersive seem to be higher. Think of AR that adds a layer of digital information over the real world, accessible through your phone or even smart glasses, as introduced through the collaboration of Meta and Ray-Ban.[4]

The first industries to make a serious move into the metaverse will likely be those already comfortable with digital transformation; tech, retail, entertainment, manufacturing, and education seem poised for early adoption. Businesses that have embraced the metaverse are already witnessing tangible benefits. For example, Shopify has started allowing merchants to create 3D models of their products, recognizing that as shopping moves into the metaverse, traditional 2D photos won't suffice. This has led to higher engagement and sales for those who have adopted the feature.

Working in the metaverse offers unparalleled advantages—everything from a dramatic reduction in business travel and associated costs, to the ability of teams to collaborate in real time on a global scale. The metaverse could make work more engaging and interactive, breaking down the barriers of geographic location and time zones. Teams can work together on a digital twin of a project (described in more detail next), making changes in real time and seeing immediate results, which in turn can significantly speed up project timelines and reduce costs.

So, while we may not all be logging in to the metaverse tomorrow, it's clear that the groundwork is being laid today for a more interconnected, immersive, and efficient future. The gradual adoption only means that we have the time to adjust, optimize, and make the metaverse a place where we all want to be.

Digital Twins

You've likely come across the term "digital twin" if you've been aware of what's going on in tech and business circles, and while it's intriguing to imagine a digital version of yourself, the reality is much broader. In simple terms, a digital twin is a virtual model of something from the real world—an object, a person, or a process. It incorporates real-time data and advanced simulations to offer a comprehensive digital view of a real-world operation. This technology is already enhancing efficiency and productivity across sectors like manufacturing, retail, telecommunications, and beyond.

For leaders, understanding the impact digital twins will have on work requires seeing their potential in how we manage and optimize operations. Digital twins allow for next-level analysis and experimentation that enable teams to test changes and optimize processes in virtual environments before implementing them in the real world. For example, you could test manufacturing operations in a virtual factory or learn how weather patterns in certain regions will impact construction. Imagine what you can do when decision-making is more informed, risks are minimized, and innovation is pursued more aggressively. Leveraging digital twins can lead to smarter, more efficient work processes that help organizations stay ahead in a competitive landscape and deliver value in new and exciting ways.

Let's consider a few examples using NVIDIA's Omniverse Enterprise, a platform that facilitates collaboration in digital twins. Powered by NVIDIA's graphics processing units (GPUs), the platform is central to the development of realistic, high-quality virtual environments.

BMW

BMW is leading the charge in automotive manufacturing by integrating NVIDIA's Omniverse into its production planning and operational systems. Within this virtual environment, an interdisciplinary team of planners, engineers, and facility managers use digital twins to model everything from machine layouts to entire factory floors. This virtual collaboration allows them to simulate intricate manufacturing processes and workflows, testing and tweaking various factors like machinery placement, robot paths, and employee workstations. What used to be a drawn-out, manual process that involved significant trial and error can now be virtually perfected before being set up in the physical world, leading to more accurate planning and fewer costly mistakes.

The impact of this technological shift has been substantial. According to BMW, leveraging the power of digital twins has resulted in a 30 percent increase in production planning efficiency.[5] This not only translates to significant cost savings but also speeds up the time-to-market for new vehicle models. The virtual environment allows for real-time adjustments and instant feedback, enabling teams to make data-driven decisions quickly. This collaborative and highly interactive approach to production planning through the metaverse is positioning BMW as a pioneer in industrial innovation, showing what's possible when cutting-edge technology is applied to traditional manufacturing processes.

Ericsson

Ericsson, a global leader in telecommunications, has embraced the use of digital twins in another fascinating way, extending the technology's applications far beyond its origins in manufacturing. Ericsson created city-scale digital twins to simulate optimal conditions for 5G antenna propagation.[6] This is a crucial application, given that the efficiency of 5G networks can be significantly impacted by a multitude of variables including terrain, architectural features, and other urban elements. Traditionally, determining the best placement for antennas required labor-intensive site surveys and testing. However, with digital twins, Ericsson can run countless scenarios in a virtual environment, effectively identifying the most advantageous locations for antennas. This not only makes the 5G networks more robust and efficient but also expedites the rollout process and significantly reduces costs.

The value of Ericsson's approach extends beyond just immediate efficiency gains. The digital twin technology allows for proactive network management by simulating how changes in the urban environment can affect network performance, enabling Ericsson to anticipate and mitigate issues like signal interference or coverage gaps. As cities evolve, the digital twin can be updated to reflect these changes, providing a mechanism for ongoing network optimization. Thus, Ericsson's use of digital twins is setting a precedent that could revolutionize 5G deployment strategies, potentially becoming a new industry standard for telecommunications companies worldwide.

Lowe's

Lowe's, known for its wide range of home improvement products, is diving headfirst into the metaverse with its Open Builder platform. Its goal is clear: democratize the metaverse for builders. The platform's motto, "We're making our pro-grade 3D products free to all, to help builders create even more possibilities in the metaverse," captures the essence of its ambition. By offering professional-grade 3D design tools for free, it's opening up a world of possibilities not just for professional contractors but also for amateur do-it-yourselfers who may have metaverse projects of their own.

Take, for example, the digital landscaping sector within the metaverse. By leveraging the Lowe's 3D products, users can virtually design backyards, complete with tool sheds, flower beds, and even intricate water features. Once satisfied, they can execute these plans in the real world, purchasing the necessary items directly through Lowe's. It's like a "try before you buy" experience but on an entirely different level.

For companies that have embraced digital twins, the benefits extend beyond improved collaboration to include enhanced customer engagement and sales performance. Research from McKinsey indicates that companies integrating digital and physical experiences through digital twins can anticipate significant boosts in their sales metrics.[7] Specifically, these companies may experience up to a 10 percent increase in overall sales revenue, coupled with a remarkable 25 percent rise in sales conversions.

This underscores the transformative potential of digital twinning not only in optimizing internal operations but also in driving tangible business outcomes and fostering deeper connections with customers. By leveraging digital twins to create immersive and personalized experiences for consumers, companies can differentiate themselves in the market, cultivate brand loyalty, and ultimately drive sustainable growth in an increasingly competitive landscape. As such, the integration of digital twinning technology represents a strategic imperative for forward-thinking organizations seeking to thrive in the digital age.

Augmented and Virtual Reality

Virtual reality has been a buzzword for years, often associated with gaming or high-tech simulations. Yet, its transformative potential for the professional world is gaining ground, particularly as we navigate an increasingly remote and decentralized work landscape. Companies like the platforms Spatial.io, Glue, Arthur, and Immersed aren't merely in the business of crafting neat tech toys; they're revolutionizing the way we interact, collaborate, and engage with our workspaces through "mixed reality workspaces."[8]

The future is more than teleporting our 2D Zoom boxes into a 3D space. It's about how these 3D spaces can augment our ability to connect, share ideas, and accomplish tasks. Imagine being on a design team spread across three continents, putting on your VR headset, and entering a shared virtual space where you can manipulate prototype designs with your colleagues in real time. This is the escape we've been needing from sending endless email threads with attachments. You're "physically" there, tweaking the design hands-on and hashing out ideas as if you're in the same room, despite being miles apart.

The adoption of AR or VR devices for collaboration brings along several advantages including the reduction of social distance, a higher satisfaction with work results, an improvement in communication, and increased comfort due to more flexibility in movement. All these factors in combination are likely to increase the adoption of the devices for work purposes in the next few years.

Also, the technologies solve the demands of the next generation of employees, enabling a more flexible workforce while maintaining closeness to colleagues. Philipp Sostmann of PwC Europe, a thought leader in immersive technologies, says, "Immersive technologies will enable for the first time in 50 years to break out of the traditional laptop or tablet display that has been limiting our interaction with digital content. For the first time we will have the freedom to design our workspace according to our needs. It will be purely task-oriented, tailored to the respective physical environment and truly human-centric. Rather than us humans adapting to the fixed laptop screen, it will be a future where technology is tailoring itself around us and our individual needs."[9]

A staggering projection from PwC estimates that the combined market for VR and AR could be worth up to $1.5 trillion by 2030.[10] This is an insight into how integral these technologies could become in the mainstream corporate milieu. Businesses are starting to look at these platforms beyond the novelty value by eyeing them as vital tools for training, collaboration, and task management. For example, Walmart has already trained over one million employees using VR, reporting a 10 to 15 percent improvement in performance.[11]

The global implications are even more profound. When remote work is quickly becoming the norm—FlexJobs reports that remote job listings increased by 52 percent in the last two years—VR's ability to closely mimic physical presence opens up a world of possibilities for talent acquisition, team cohesion, and even workplace well-being, but also fosters a sense of belonging by creating immersive and inclusive environments that transcend geographical boundaries.[12]

AR and VR were once high-tech distractions or a futurist's dream. Not anymore. They're an emerging pillar of the new world of work, where their contributions go far beyond making meetings more interactive. They have the potential to redefine the meaning of workplace presence, shaping a future where distance is no longer a barrier to full engagement. The future of work is not about where we are but what we can achieve, irrespective of location.

The Immersive Internet for Employee Experience

For you, as a corporate leader, the immersive internet is a revolutionary tool for transforming how you engage your workforce. From streamlining onboarding processes to fostering innovation faster, cheaper, and more creatively than before, this technology is all about boosting productivity and making employees feel more deeply connected and engaged in their work. Consider Accenture's Nth Floor.

Accenture's Nth Floor

Accenture's Nth Floor is heralded as the world's largest enterprise metaverse. It seamlessly integrates elements of onboarding, immersive learning, collaboration, social interaction, and wellness, marking a monumental shift in the employee experience. Beyond revolutionizing the onboarding journey for new hires, this cutting-edge platform confronts critical challenges such as hybrid working arrangements and the global enhancement of leadership skills.

Olly Jeffers, Accenture's global onboarding metaverse lead, shares insights into the platform's evolution. What initially began as an onboarding platform for 150,000 new hires, offering a virtual tour of Accenture's global offices, has blossomed into One Accenture Park—a vibrant hub that welcomes all employees. Here, individuals are immersed in networking opportunities, peer interactions, and the rich tapestry of Accenture's culture and ethos. The journey commences on their second day, as Jeffers elaborates on the two-part experience: "Initially, new joiners find themselves in a meticulously crafted virtual conference room, mirroring the physical ambiance of Accenture's offices. Here, they mingle with fellow newcomers, acquaint themselves with the technology, and begin to forge connections. Subsequently, their 'park ranger' guides them through a corridor, unveiling One Accenture Park in a cinematic Jurassic Park-style."

Post-onboarding surveys reveal an overwhelming 94 percent favorability rating for One Accenture Park. Delvin Monzon, a technology architect at Accenture, shares, "The inclusion of One Accenture Park offered a refreshing approach to learning and interaction within my new start group. Exploring the metaverse and engaging with VR technology injected an extra layer of excitement and enthusiasm. We delved into virtual socialization, uncovering features like exploring different Accenture offices, projecting presentations on skyscrapers, and indulging in interactive games."[13]

This hands-on approach provides new joiners with a glimpse into Accenture's work culture while simultaneously boosting engagement and eliminates distractions typically associated with remote onboarding. What truly sets One Accenture Park apart, however, is its unexpected evolution. "We discovered that the metaverse provided our people with a unique social and collaborative outlet—an avenue for spontaneous connection during times when physical gatherings were restricted," explains Jeffers. This realization led to the creation of One Accenture Park—a haven for new joiners and a testament to Accenture's commitment to fostering an enriching employee experience.

Accenture has orchestrated numerous community gatherings on the platform, including an annual strategy meeting attended by 150 managing directors from 25 countries. According to Monzon, "Participants then transition into smaller groups for bonding exercises before delving into immersive learning sessions. Guided by a mock client, they tackle real-world business challenges as a team."

By seamlessly blending technology with human connection, Accenture's Nth Floor continues to redefine the boundaries of employee engagement and collaboration in the digital age. Within Accenture's Nth Floor metaverse, a virtual campus called One Accenture Park is helping new employees personally connect with the culture and each other. Overall, Accenture's Nth Floor represents a bold experiment in reimagining the traditional office space for the digital age, demonstrating how companies can leverage immersive technologies to create more engaging and collaborative work environments.

The Metaverse for the Corporate Leader

When it comes to captivating your workforce, especially when vying for the attention of younger talent, delving into immersive internet technologies opens a realm of exciting possibilities. Here are just a few examples of what these cutting-edge technologies can bring to the table:

> *Teamwork unleashed.* Remember when Nike created a virtual running experience in which users could race against their friends anywhere in the world? With the immersive internet, your teams can collaborate in vibrant virtual environments, sparking creativity and innovation regardless of where they're located.

> *Get ready for wow.* Take a page from Walmart's book, which used VR to train employees in real-life scenarios like dealing with holiday rush crowds. Embrace VR and AR to create immersive experiences that'll have your employees buzzing with excitement, making training sessions feel more like epic

adventures and onboarding experiences that leave a lasting impression.

Personalized workspaces. Imagine if your employees could design their dream work environments. That's exactly what Accenture did when it created a virtual office space where employees could customize their workspace to suit their preferences. Flexible and customizable virtual offices cater to different work styles, boosting morale and productivity.

Connect like never before. Dive into immersive conversations as Mozilla did when it launched Hubs, a platform that allows teams to meet and collaborate in VR. Communication goes beyond boring video calls, making everyone feel included and engaged, no matter where they are.

Fuel creativity. Audi used VR to create a virtual showroom where customers could customize and interact with their dream cars. Give your team the tools it needs to bring its wildest ideas to life, driving your business forward.

Tackle challenges together. There are hurdles like tech complexity and security concerns. But with the right approach and collaboration, you can overcome them, just like companies such as IBM and Microsoft are doing as they navigate the complexities of virtual worlds.

Embrace the immersive internet and your workforce will become more engaged, productive, and passionate about driving your business to new heights. For traditional workers, the metaverse brings opportunities to enrich job roles, expand skill sets, and even reimagine career trajectories. It offers avenues for real-time collaboration and problem-solving that are unbound by geographical or organizational constraints.

The changes driven by the immersive internet are already unfolding, whether we're ready or not. To merely observe from the sidelines is to forfeit a role in a future that promises a radical transformation of what we consider work to be. The best time to prepare and become an active participant in this transformation is right now. So, seize the reins and become an architect of this new professional universe.

CONVERSATION STARTERS

1. Considering the potential of the metaverse to reshape remote teamwork, what strategies can you develop to leverage virtual spaces for enhancing creativity, collaboration, and a sense of presence among remote teams?
2. How can your organization implement digital twins to simulate, analyze, and optimize your operations, whether in manufacturing, product development, or service delivery? In what ways can digital twins contribute to predictive maintenance, performance optimization, and decision-making processes to drive efficiency and innovation?
3. Reflect on the potential of AR and VR technologies to revolutionize employee training, customer engagement, and product demonstrations. How can immersive experiences be designed to not only enhance learning outcomes but also build deeper emotional connections with brands and products?
4. As digital and physical workspaces converge, what challenges do you anticipate in managing a workforce that operates within these blended environments?
5. What leadership qualities and strategies will be crucial for navigating the future in the metaverse?

PART III

Future of Work3

CHAPTER 6

The Rise of Decentralized Autonomous Organizations (DAOs)

In political philosophy, the concept of "consent of the governed" refers to the idea that a government's legitimacy and moral right to use state power is justified and lawful only when the people or society over which that political power is exercised consent to it. It's this philosophy that drove Kevin Owocki to organize his company into a decentralized autonomous organization (DAO) rather than an LLC. He observed the foundational role of open source software and the inefficiencies in traditional recruitment processes and was driven to seek a more inclusive and efficient solution. Thus, Gitcoin was born, a platform designed to fund open source projects through a decentralized network by harnessing the collective power and wisdom of its community. "We didn't set out to create just another web2 startup," Owocki remarked in a post about Gitcoin's history.[1] "The grand vision behind launching a DAO was to bring Gitcoin under the governance of its very own users."

This vision emphasizes a trajectory of sustained growth and decentralization to amplify the platform's impact within the open source and blockchain realms. But the implications of Gitcoin's shift toward a DAO model extend far beyond its immediate community.

Gitcoin embodies a departure from conventional corporate frameworks by facilitating a work culture steeped in shared governance. Through the innovative use of blockchain technology, Gitcoin has orchestrated a system that acknowledges and rewards individual contributions and democratizes governance, and the collective intelligence of its community informs the strategic direction. This empowers Gitcoin to pursue its goals with greater agility and community alignment, which presents a compelling blueprint for how organizations might operate in the future.

In this chapter, we explore a vision of the future in which work transcends being solely a means to an end to become a participatory and enriching journey where every individual contributes to building something far greater than themselves. This transformative approach to work is made possible through the innovative structure of DAOs.

How Do They Work?

DAOs are entities that operate without centralized control, governed by smart contracts and consensus among its members rather than by a centralized hierarchy. A DAO is a revolutionary concept sprung up within the fertile ground of blockchain and cryptocurrency technology that aims to break free from traditional centralized structures. The idea behind DAOs is to create a world where organizations run themselves, independent of human management, guided solely by lines of code and the collective wisdom of their participants. They represent the dream of creating self-governing entities that function autonomously. This type of organization operates without hierarchy, where decisions are made by consensus, rules are hard-coded into inviolable smart contracts, and anyone can be an owner.

Imagine a workplace where your boss isn't a person but a set of smart contracts. You wake up in the morning, open your laptop, and consult a digital interface. Here, you are part of a decentralized, self-governing collective making decisions on everything from budget allocations to what projects to take on.

A DAO is represented by rules encoded as a computer program that is transparent, controlled by the members of the organization, and not influenced by a central body. DAOs are the most effective way of establishing a digital company and are typically set up to run on blockchain technology, especially Ethereum. They embody the principles of web3—decentralization, transparency, and peer-to-peer interactions. Here's what DAO governance can look like:

Making proposals. If you have an idea, like funding a new project or changing some rules within the organization, you can submit your proposal to the DAO's community members.

Everyone gets a say. Each member of the DAO has a voice in decisions. The more tokens you hold, which are distributed on the Ethereum blockchain, the more influence your vote carries. A digital voting system is executed on the blockchain through platforms like Snapshot.

Smart contracts do the work. Once the votes are in and a proposal meets the required support, there's no need for a manager or human intervention. Smart contracts, which are like self-executing computer programs, take care of executing the decision.

Full transparency. Every step of the process, from submitting proposals to the final decision and all the transactions, is recorded on the blockchain. This way, everyone can see what's happening, ensuring transparency and accountability.

Code rules. DAOs are ruled by code. The initial code defines how they work, and members have to approve any changes to that code through the DAO's own decision-making process, keeping things democratic and avoiding central control.

These digital-first entities challenge how we think about work and how we conceive of organizations themselves. It's important, then, to ask, Why are DAOs crucial in the context of a changing workplace? And more importantly, what does their advent mean for leaders navigating the complexities of the emerging workforce?

Understanding a DAO's Structure and Functionality

To fully understand DAOs and how they operate, you need to understand how they're composed (see figure 6-1).

It should comfort traditionalists that our current notion of full-time employees will still exist in a DAO, whom the industry is currently calling "core contributors" (see the center of the figure). They are a small subset of individuals deeply invested in a single project or DAO, much like a core team in a traditional corporate setup. However, these teams will be leaner and can complete much more work by leveraging AI and automation. The early stages of Instagram are an example; it had a small team of thirteen when acquired by Facebook for $1 billion.[2] Small teams making a big impact will become the new norm.

FIGURE 6-1

Anatomy of a DAO

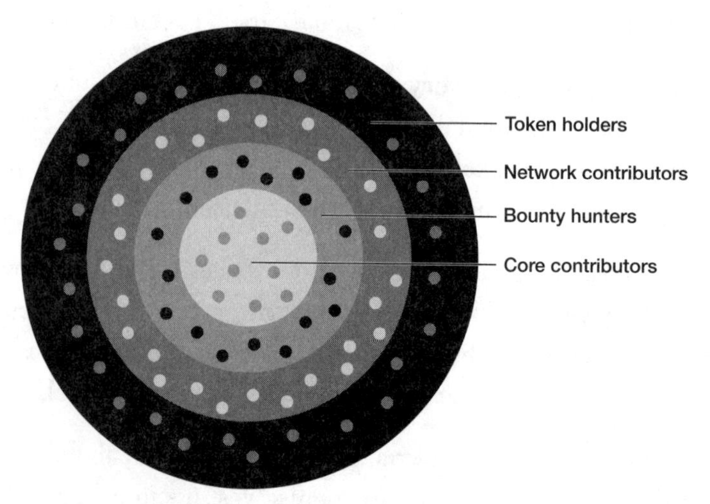

Graphic inspiration from Brian Flynn, Zakku, and the Orbit team.

The next layer will be filled by "bounty hunters," or specialists who work on specific tasks for different DAOs. Unlike traditional freelancers, they operate in a more transparent and collaborative environment, thanks to smart contracts and decentralized decision-making. Some even group together to form services DAOs.[3] For example, Llama is a service DAO specialized in treasury management; RaidGuild offers software development services. Bounty hunters are the gears in the decentralized machine that handle specialized tasks that a small core team can't manage alone.

The most exciting and transformative part of the system is the layer of "network participants." Think of these as the everyday users who contribute value just by participating in the network, like playing games or testing the user experience to provide data quickly. This model is called "X-to-earn," for example, play-to-earn, learn-to-earn, create-to-earn, and so on. For example, Decentraland is a blockchain-based virtual reality world where players can buy, develop, and monetize virtual land using the platform's native cryptocurrency, MANA. Players can build structures, create artwork, and even host events on their virtual properties, and they can earn real-world income through these activities. The game has gained significant traction, with land sales sometimes reaching several thousand dollars, showing another angle of how the play-to-earn model can be both rewarding for users and profitable as a business.

In the world of DAOs, where activities are more transparently rewarded, network participants could soon be a new layer of your workforce ecosystem. This flexible and inclusive work landscape is what sets DAOs apart from traditional organizational models. For instance, community members in Sandbox, a decentralized virtual gaming world, can earn rewards by contributing to the platform's development, from designing virtual real estate to crafting in-game experiences. These rewards are not just tokens but represent real financial value and decision-making power within the community. This mechanism facilitates a democratic and meritocratic work environment, demonstrating how DAOs like Sandbox are revolutionizing the concept of employment.

DAOs are redefining the "why" behind work by making governance an active, on-chain participation rather than just an idea. Consider nounsDAO, where much of the governance happens on-chain, fully embracing the decentralization ethos. This setup transforms work from being a route to a paycheck into a purpose-driven contribution to a community. Through on-chain voting and discussions, members are directly involved in shaping the DAO's direction, turning work into an act of community participation.

These layers of the DAO ecosystem collectively contribute to a new, dynamic model of work that not only enhances efficiency and productivity but allows every participant to have a stake in the outcome. The shift toward such a decentralized, inclusive work environment underscores the transformative potential of DAOs beyond traditional organizational structures. Now let's review a real-world example.

A DAO at Work

From its early days, MakerDAO, a decentralized lending platform, navigated the choppy waters of regulatory scrutiny, technical challenges, and the inherent skepticism surrounding the feasibility of a stable and decentralized currency. Despite these hurdles, MakerDAO thrived, bolstered by a community of believers united in their quest for a more inclusive and transparent financial system. The key to its success lies in its governance model: a system where decisions are made not by a select few in boardrooms but by a global community of MKR token holders through a transparent voting process. MKR acts as a governance and utility token to the Maker Protocol with no fixed supply.

This model has been tested time and again. MakerDAO's transition to Multi-Collateral DAI, its internal cryptocurrency, in 2019, for instance, was a litmus test for the community's ability to steer the DAO through significant changes. And steer it did, with the kind of collective wisdom and agility seldom seen in traditional corporate structures. Each challenge surmounted has been a testament to the resilience of the DAO model, proving that decentralized governance can indeed match, if not surpass, the effectiveness of its centralized counterparts.

In the DAO, contributors are motivated not just by financial incentives but by the shared vision and values of the community. Employment in such a context transcends the transactional nature of traditional job roles by offering a sense of purpose and belonging that is often missing in conventional workplaces. MakerDAO's decentralized model supercharges its culture to be innovative and agile, where contributions are value-based on merit rather than hierarchy.

The narrative of MakerDAO is a blueprint for a new way of working and organizing. It highlights the possibilities of collective decision-making, community engagement, and a workforce unified not by geography or corporate structure but by shared goals and ideals.

In 2021, Wyoming became the first state to pass legislation designating legal entity status to DAOs by creating a DAO LLC statute. As of this writing, four other states have followed suit. DAO incorporations are also available globally including in the Marshall Islands, Switzerland, the Cayman Islands, Liechtenstein, Singapore, Panama, Gibraltar, and the Bahamas. While DAOs aren't yet common, they appear to be gaining traction in many optimistic states, countries, and communities.

The Shortcomings of a DAO

While there is promise in a more decentralized way of operating, leaders venturing into the realm of DAOs must be acutely aware of the potential pitfalls that accompany the decentralized model. For instance, when "The DAO"(as it was aptly named) burst onto the scene in 2016, it ignited imaginations and set the stage for a new era of decentralized innovation. The organizers of The DAO aspired to be a new kind of venture capital fund, sans traditional management structure or board of directors. Rather, The DAO operated on open source code and aimed to revolutionize the way both commercial and nonprofit enterprises were organized. Within twenty-eight days, The DAO managed to capture widespread interest among enthusiasts and amassed a staggering $150 million through crowdfunding.[4]

But the cracks in this revolutionary idea were exposed when an anonymous hacker exploited a vulnerability in The DAO's code, draining

$50 million worth of ETH tokens, the Ethereum blockchain's native cryptocurrency. The hack had far-reaching implications, considering that The DAO, at the time, held a remarkable 15 percent of all ETH in existence.[5] The value of ETH plummeted, from over $20 per token to below $13.

So the Ethereum community faced a divisive decision: allow the theft to stand and let the thief get away with $50 million, or intervene to reverse it to return the money to the investors. Eventually, it opted for a controversial "hard fork," essentially rewriting the blockchain to return stolen funds to the original investors. This decision effectively split Ethereum into two separate blockchains: Ethereum and Ethereum Classic. Because of this incident, public perception of web3 technologies and cryptocurrencies took a significant hit. And critics lambasted the field, labeling it a scam, a Ponzi scheme, and an exercise in "pulling money out of thin air."[6] The incident illuminates critical vulnerabilities, from technical security flaws to complex ethical and governance challenges. Leaders should understand that while DAOs can democratize decision-making and spur innovation, they also require a nuanced approach to governance that balances collective input with effective action.

While DAOs offer a compelling vision for the future of work, it's important to temper the optimism with a dose of reality. The use of digital currency, smart contracts, and tokens may align incentives and facilitate decentralized governance, but these systems are far from foolproof. DAOs, after all, are people-based organizations, and wherever there are people, there's potential for politics, corruption, and other challenges.

The technology that makes DAOs promising also exposes them to risks like hacks and security breaches. Governance is often cited as a powerful feature of DAOs, but it's still in a nascent stage. Despite platforms like Snapshot and Orca aiding in the process, decision-making can become convoluted, leading to voter apathy or concentration of power among a few. Come to find out, democracy is hard.

Managing a DAO's treasury and allocating resources efficiently also present significant challenges. Tools such as Coordinape and

SourceCred have been developed to democratize these processes, allowing team members to have a say in pay decisions and contributions recognition. However, the systems come with their own complications. Without centralized oversight, there's a heightened risk that incentives may not always align with the organization's broader goals. Additionally, favoritism and the possibility of individuals manipulating the system for personal gain could undermine fairness and effectiveness. These issues underscore the need for careful design and constant vigilance in decentralized financial management practices.

Though the sector is growing—as evidenced by a 151 percent year-over-year increase in full-time roles in blockchain and crypto, according to a Blockchain Academy report—sustainable income via DAOs remains an uncertain proposition.[7] The idea of "X-to-earn" is alluring but largely experimental. It's not a panacea for the long-standing problems of work, and we're far from seeing DAOs as stable, long-term employment options for the masses.

To quote author William Gibson, "The future is already here, it's just not evenly distributed."[8] DAOs are a part of that unevenly distributed future. As a leader, you will undeniably have to navigate the potential for DAOs to revolutionize work, yet the challenges inherent in these systems highlight the need for robust strategies and oversight. You must recognize that while DAOs represent a significant shift toward more democratic and transparent work environments, they also require a diligent focus on governance, ethical practices, and safeguarding against manipulation. Indeed, the journey into DAOs as a sustainable future of work is filled with both opportunities and obstacles.

The Future Is DAO

That brings us to the implications for job roles and career paths. In a DAO, contributors are not climbing a corporate ladder; they are navigating a network. Their value isn't determined by a title but by their contributions, recognized and rewarded through tokens that offer a stake in the organization's success. This could be liberating, but it could also be disorienting for those accustomed to traditional career trajectories.

In a traditional corporate structure, managers are the linchpin around which many critical decisions are made. They allocate resources, oversee projects, resolve conflicts, and are a conduit for information flowing between employees and executives. But within the architecture of a DAO, the role of a manager as we know it becomes largely redundant. The entire concept of "management" undergoes a strategic makeover.

This shift invites the question: In a DAO, is there a role left for managers? Perhaps, but not in a form we currently recognize. Managers may morph into facilitators, advisers, or specialized contributors rather than decision-making authorities. In a DAO, the community leads, and the so-called manager listens—a paradigm shift, but a welcome one for many.

DAOs are a social experiment that could redefine how we think about organizations and work. As leaders, it's your responsibility to understand this emerging landscape, not just to remain competitive but to create a more democratic future for work. It's also your role to ask the tough questions: How do we ensure that the democratic approach of a DAO doesn't lead to decision-making paralysis or abuse? How do we maintain accountability in a system where authority is diffused?

These are the questions that digital-first leaders are grappling with, and they are finding unique solutions, like setting quorums for different types of decisions and implementing tiered voting systems that streamline the efforts of their team. They are using emerging technologies to ensure that contributions and decisions are traceable and visible to all members. They are implementing smart contracts for task and project management that automates and enforces accountability. This transformation is driven by the desire to address long-standing organizational challenges in a novel and effective way.

CONVERSATION STARTERS

1. What are the key motivations behind organizations transitioning to a DAO model, and how might these motivations align with your company's vision and values?
2. How do the operational mechanisms of DAOs, such as making proposals, voting, and executing decisions through smart contracts, challenge your existing decision-making processes? How can you integrate these mechanisms into your organization?
3. How can you incorporate transparency and accountability into your current operations to build trust and foster a more open work environment?
4. How can your organization prepare for a shift to remove traditional managerial roles and hierarchies, and what new roles or skills should you prioritize to support this transition?
5. What are your strategies for addressing potential pitfalls, such as decision-making paralysis, security vulnerabilities, and maintaining a coherent company culture in a decentralized environment?

CHAPTER 7

Smart Contracts and Self-Executing Agreements

One of the first recorded instances of a handshake appears in a ninth-century BC stone relief featuring Assyrian King Shalmaneser III clasping hands with a King Marduk-Zakir-Shumi I of Babylonia to cement a partnership and convey peaceful intentions among armed men.[1] The revered poet Homer frequently mentioned handshakes in his epic works *The Iliad* and *The Odyssey*, usually as symbols of vows or trust for an exchange of goods and services. These hand-clasping gestures reverberate through the annals of history, from Greek funerary art to Roman coins, symbolizing collaboration, loyalty, and agreement.

As societies developed and legalities became more complex, the humble handshake underwent radical transformation. First, it was supplemented with written contracts and then, as we progressed into a digital world, a contract with electronic signatures. More recently, we've begun using cryptographic hash codes, or a unique digital fingerprint that ensures the data integrity of transactions. Here's how they relate: these changes weren't simply bureaucratic embellishments; each shift aimed to remedy the shortcomings of its antecedents. Digital signatures, despite their convenience and legality, still depend on manual verification and third-party facilitators.

The introduction of smart contracts represents a watershed moment in the evolution of agreements. These innovations go far beyond digital convenience and will fundamentally reshape how we establish trust and how we execute contracts. By automating contractual obligations and removing the need for third-party verification, smart contracts make transactions faster, more secure, and less susceptible to human error or malfeasance. As a result, they stand to revolutionize various sectors such as real estate, health care, and finance, offering a glimpse into a future where trust is encoded in lines of code and the act of making an agreement is as streamlined as it is secure.

Smart Contracts: The Basics of Trustless Trust

You've likely heard the phrase "trust but verify." In a decentralized gig economy, it's more like "just verify." Thanks to blockchain technology, every transaction, review, and work assignment is publicly recorded and immutable. Trust is coded into the system itself.

Imagine finding your dream home online, agreeing on a price with the seller, and completing the purchase—all without the mountain of paperwork, weeks of back-and-forth negotiations, or fear of the deal falling through at the last minute. That's the streamlined experience smart contracts aim to bring to the real estate world. With smart contracts, every step, from agreeing on terms to transferring ownership, can be automated and secured, making the buying process as simple and stress-free as unlocking your new front door (see figure 7-1).

What makes the concept of smart contracts so game-changing is the notion of *trustless trust.*[2] This might sound like an oxymoron, but it is actually a revolutionary concept, especially when talking about contracts and agreements. Traditionally, when you enter into a contract with someone, you need to either trust them personally or trust a legal system to enforce the agreement if things go wrong. This can be slow, costly, and sometimes unreliable.

Trustless trust, on the other hand, relies on mathematics and computer code instead of people or institutions and is the foundation of

FIGURE 7-1

Comparison: Traditional contract versus smart contract

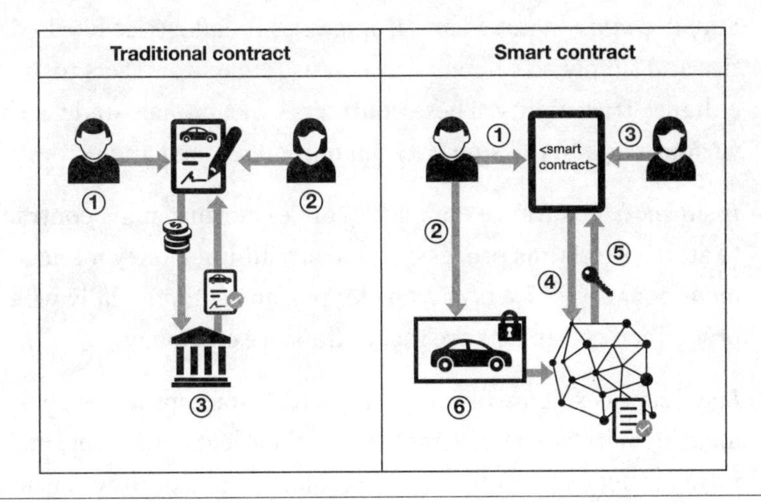

smart contracts. At its core, a smart contract is an automated, self-enforcing agreement where the contract terms are directly coded into a blockchain. It's a digital contract that spells out the terms of an agreement and actively enforces them without human intervention. The beauty of this system lies in its automation and impartiality. When a smart contract is deployed, it performs its obligations automatically once the conditions coded into it are met, without the need for human intervention or oversight.

This shift is tremendous because it streamlines transactions to make them faster and more efficient. There's no need to wait for a third party to confirm the deal or resolve disputes, no legal team to verify or a third party to facilitate the transaction, which can save a lot of time and resources. In a world where trust is hard to come by, the trustless trust of smart contracts offers a secure and reliable way to conduct transactions and agreements, powered by the impartiality and reliability of technology. Trust becomes an asset, hard-coded into the very code of business interactions.

A number of industries have already begun using, or have shown interest in using, smart contracts:

Supply chain management. Companies in industries like logistics and supply chain management use smart contracts to enhance traceability. These contracts can automate and verify various steps in the supply chain to reduce fraud and errors.

Insurance. Insurance companies are exploring smart contracts to automate claims processing and streamline policy management. Smart contracts can trigger payouts automatically when predefined conditions are met to improve efficiency.

Legal services. Law firms and legal tech startups are experimenting with smart contracts for various legal purposes, including contract execution, dispute resolution, and escrow services.

Health care. The health-care industry is exploring smart contracts for patient data management, clinical trials, and medical billing. These contracts can help secure and automate sensitive processes.

Financial services. Beyond decentralized finance (DeFi), traditional financial institutions are also looking into smart contracts to automate settlement processes, trade finance, and compliance tasks.

Government. Some governments are exploring the use of blockchain and smart contracts for transparent and secure voting systems, identity verification, and public record-keeping.

Energy and utilities. Energy companies are using smart contracts for peer-to-peer energy trading and grid management. These contracts enable consumers to buy and sell excess energy directly.

Tokenization of assets. Companies in various industries are exploring the tokenization of assets such as art, real estate, and commodities. Smart contracts are used to represent and manage these digital assets.

As innovative as these smart contracts are, the opportunity goes well beyond the industries in these examples. Many companies are already harnessing smart contracts for various operational facets, ranging from procurement to copyright management.

How Smart Contracts Could Transform the Workplace

Smart contracts offer potential solutions that could dismantle long-standing obstacles to efficiency within HR practices. Consider the example of Talentpair. Talentpair is an award-winning trailblazer in the recruitment landscape that accelerates the hiring process through state-of-the-art AI algorithms. Recognizing the inefficiencies and lack of trust that have plagued the recruiting industry for years, Talentpair made a strategic shift to a decentralized model in recent years, further strengthening its position as an innovative leader in the space.

The company first introduced a concept it calls "non-fungible talent," a play on the concept of non-fungible tokens (NFTs), unique digital identifiers recorded on a blockchain and used to certify ownership and authenticity. Non-fungible talent, then, is the creation of a unique digital résumé for each worker on the blockchain, which AI then utilizes to more effectively match workers with suitable job opportunities. This approach eliminates the reliance on self-reported résumés and blind trust to ensure all credentials are transparently and securely recorded.

What makes Talentpair truly innovative, however, lies in its use of smart contracts. In our interview with Joe Kosakowski, president and founder of Talentpair, he said, "Smart contracts offer unprecedented levels of capacity and fair compensation that benefits everyone involved in the recruitment life cycle." He illustrated this with an example: within traditional recruiting models, if you introduce someone from your network and another recruiter helps them secure a job, the commission typically is given to the recruiter who closed the deal, but with a smart contract, the commission can now be distributed fairly among all contributing recruiters. In this scenario, the person making the initial introduction would earn a percentage of the commission, which is automatically distributed the minute the smart contract is fulfilled.

This process ensures that each recruiter's contribution is recognized and rewarded without fear that you'll be undercompensated.

This revolutionary approach has not gone unnoticed, as Talentpair was named the 2023 RemoteTech Overall Recruiting Solution of the Year, marking its third consecutive year winning a top solution award.[3] By incorporating smart contracts into its operational DNA, Talentpair has effectively dismantled the constraints that have long plagued traditional recruiting methods. No longer do recruiters and companies have to wrestle with onerous administrative tasks, nor do they have to fret about compliance issues. With smart contracts, actions are automated and rewards are instantaneously dispensed, all recorded transparently and fairly for whoever participated in supporting the hiring of a candidate.

Talentpair's pioneering use of smart contracts highlights the future of hiring and indicates a larger shift toward a more equitable and accountable professional ecosystem. We can begin to see how these digital contracts could redefine the landscape of employment, from recruitment and onboarding to performance management and dispute resolution. Let's explore the transformative potential of smart contracts in reshaping HR and employment specifically.

Smoother payment processes. With smart contracts, the once cumbersome process of ensuring that employees are paid on time and accurately becomes effortless, their earnings automatically adjusted and disbursed based on predefined criteria. This extends to managing benefits, where the allocation to health insurance or retirement plans is no longer a manual task but a dynamic, real-time adjustment to the changing needs of the workforce. Companies with external collaborators including subcontractors, freelancers, app developers, marketplace sellers, and so on face the uncertainty of payment and the shadow of disputes. Smart contracts illuminate this space with the promise of fairness and prompt compensation, directly releasing funds upon the completion of agreed milestones, and thereby fostering a foundation of trust between parties previously bridged by skepticism.

Security and confidentiality. The immutable and secure nature of blockchain guarantees that sensitive agreements are preserved against tampering, accessible only to those who hold the keys. This security is paramount in an era when data breaches and privacy concerns loom large over corporate and personal interactions alike.

Speed and efficiency. Project management, too, finds an ally in smart contracts. The life cycle of a project, punctuated by approvals, payments, and the initiation of subsequent phases, can be streamlined, reducing the administrative overhead and propelling projects forward with unprecedented speed and accuracy.

Streamlined dispute resolution. Disputes are an inherent aspect of human relationships and interactions in the workplace and often arise from misunderstandings or disagreements over terms and conditions. Smart contracts offer a groundbreaking solution by acting as a neutral intermediary and executing decisions impartially based solely on the logic and conditions mutually agreed on by all parties involved. This could minimize the potential for conflict and establish a transparent and straightforward process for resolving disputes to foster a more harmonious workplace environment.

Enhanced skills and professional development. Achievements in training and certifications can trigger recognition and rewards automatically, thereby encouraging employees' continuous learning and growth. By tracking progress and verifying achievements in real time, HR creates a culture where growth is both encouraged and visibly appreciated. This innovation leads to an upskilled workforce that enhances individual and organizational success.

Now that we've sketched out the expansive potential of smart contracts and how they can reshape organizational dynamics, let's shift our attention to the foundational mechanics that enable these innovations.

The Mechanics of Smart Contracts in Employment

Let's say you're a business owner in need of a freelance graphic designer for a two-week project. Using the old way of hiring, you'd create a contract outlining terms, deliverables, and payment schedules.

This document would go back and forth between HR and legal teams before even reaching the designer. Once they complete the work and it gets approved, you still have to wait for finance to release payment.

Now, self-executing smart contracts with the terms directly written into lines of code automate the entire process. When the freelance graphic designer uploads the final design and it gets approved—either manually or through a set of predefined conditions—the smart contract automatically releases the payment. There's no waiting for checks to clear or worrying about disputes over deliverables; it's all programmed into the contract.

This is where criticism often surfaces: "But someone still has to approve it, right? Especially with something that's subjective, someone has to say the job was done to the quality expected." For tasks that inherently carry subjective quality assessments, such as creative or judgment-based work, the approval process doesn't necessarily become obsolete; rather, it evolves. Approval may come from the client through manual confirmation in a way that reflects traditional feedback loops. Alternatively, for more objective criteria, the process could leverage a set of predefined conditions that both parties agree on beforehand, potentially incorporating community validation or the use of decentralized oracles, a set of networks that enable the creation of hybrid smart contracts.

There are potential scenarios in which adaptive smart contracts could be tailored to both employer and employee needs. For instance, if you are employed on a project basis, your smart contract could automatically adjust your remuneration based on the timeliness and quality of your work. Submit high-quality projects ahead of schedule, and the contract increases your payment rate. Conversely, deliver subpar work or miss deadlines, and you could see a dip in your pay. This automation aims to maintain fairness and quality assurance to ensure that the work meets the agreed-on standards before the smart contract executes the payment.

One of the standout applications of smart contracts within HR is the automation of performance-based incentives. Traditional methods often involve time-consuming verification processes and the potential for human error, leading to delays in recognition and reward for employees' efforts. However, smart contracts are programmed to automatically recognize when an employee meets certain performance milestones, instantly triggering the disbursement of bonuses or other incentives.[4] This accelerates the reward process and eliminates discrepancies to safeguard fair and timely compensation.

The utilization of smart contracts extends beyond financial incentives to encompass various aspects of employment contracts, such as onboarding, benefits administration, and compliance with workplace policies. For example, agreements regarding leave policies, health-care benefits, and other employment terms can be encoded into smart contracts. This ensures that both parties—the employer and the employee—adhere to the agreed terms, with the blockchain providing a tamper-proof record of all transactions and interactions.

Such a high degree of automation and transparency fosters a workplace culture that values accountability and trust. Employees can have confidence in the fairness and impartiality of the systems governing their rewards and employment conditions, knowing that all parties are held to the same standards. For HR professionals and leaders, the appeal of smart contracts lies in their potential to reduce administrative burdens that allows them to focus on more strategic initiatives like talent development and employee engagement.

As organizations lead the way in integrating smart contracts into their HR practices, it's clear that this technology offers more than just operational efficiencies. It represents a shift toward more transparent and equitable employment structures, suggesting a future where the relationship between employers and employees is fundamentally enhanced. Understanding and adopting smart contracts could be key to navigating the future of work, presenting an opportunity to redefine labor relationships and establish a new standard for the employment landscape globally. (For more, see the sidebar "Enhancing Worker Stability with Smart Contracts.")

Enhancing Worker Stability with Smart Contracts

After years of waning influence, labor unions are making a remarkable comeback in the United States. From October 2022 to March 2024, there was a 57 percent surge in union representation petitions filed with the National Labor Relations Board, compared with the same period the previous year.[5] Public support for unions is also at a historic high, with a recent Gallup poll showing that 68 percent of Americans are in favor—the highest level since 1965.[6]

Even major corporations, previously untouched by unionization, are feeling the shift. In 2022, employees at fifty-four Starbucks locations voted to unionize, citing issues like poor treatment, inadequate pay, and insufficient training.[7] An Amazon warehouse in New York City made headlines by voting to establish the company's first labor union.[8] Apple employees are also gearing up for a unionization vote in early June 2024. This resurgence is driven by workers seeking better pay, benefits, and safer working conditions, and it's reshaping the landscape of labor relations in the United States.

Historically, union negotiations have been burdened by complex discussions and reams of paperwork. Imagine a scenario in which these protracted negotiations could be automated and enforced

Challenges and Ethical Considerations

Yes, smart contracts offer many advantages; however, this promising space is not without its pitfalls. Challenges such as potential coding errors, legal recognition, and the need for a robust regulatory framework make equitable and secure implementation difficult. The nascent technology comes with inherent risks such as vulnerabilities in smart contract coding, which can be susceptible to hacks, as well as the overarching specter of regulatory oversight that may impose new constraints or trigger legal consequences. Added to this is the volatility often associated with crypto assets and the still-developing infrastructure that supports them.

While the phrase "legal recognition of smart contracts" captures a slice of the challenge, it barely scratches the surface of more intricate in real time through smart contracts. Such contracts could delineate everything from work hours to scheduled wage increases, and thanks to blockchain technology, they would be both transparent and immutable.

The remarkable increase in union representation petitions and the growing public support underscore a collective drive toward enhancing work conditions, fair pay, and benefits. As we witness traditional labor forces embracing modern technology, the potential of blockchain and smart contracts to streamline and secure union negotiations emerges as a transformative tool. By automating complex negotiations and ensuring transparency and immutability, blockchain technology promises not only to reduce the costs and time associated with these processes but also to reinforce the principles of fairness and equity in the workplace. This integration of cutting-edge technology with traditional labor movements provides a glimpse into a future where labor rights are protected with unprecedented efficiency and integrity, marking a significant leap forward in the quest for better work environments and labor relations.

---

issues. One of the most pressing concerns revolves around liability. Who is responsible for the actions triggered by a smart contract? This dilemma is apparent in the case of Tornado Cash, a cryptocurrency mixing service that came under legal scrutiny.[9] The developers of Tornado Cash were indicted for allegedly facilitating money laundering, spotlighting the murky waters of ethical and legal responsibilities in the DeFi space. The smart-contract technology underlying Tornado Cash autonomously processed transactions, obscuring the origins of cryptocurrency assets to enhance privacy. However, this same feature purportedly made it a tool for laundering illicit funds.

Even if two parties agree to the terms of a smart contract within a traditional legal framework, what about accountability? If something goes awry—illegal activities, coding errors, or even unforeseen consequences that change the nature of the project—pinpointing responsibility becomes convoluted. In the virtual realm, jurisdiction itself becomes nebulous, as it's unclear where the action of a smart contract technically takes place, which complicates efforts to bring any liable parties to account.

For example, let's say that Janet, a brilliant software engineer, is bound by a smart contract to produce a set number of lines of code per day. The smart contract is programmed to measure her output and flag her for underperformance if she doesn't meet the quota. When Janet suddenly faces a medical emergency and can't meet her daily quotas, the automated and immutable nature of the smart contract fails to accommodate her human condition, leading to financial loss and potential damage to her professional reputation. But the complexities don't end there. What if Janet produces code, but it contains errors? What if the code is error-free but doesn't accomplish the intended task? What if she meets her quota, but later it's discovered that the project specifications were incorrect, rendering her efforts counterproductive? Each of these scenarios underscores the limitations of a rigid, automated system in capturing the nuances of human work and ethical considerations. Janet's example also highlights the difficulty of mixing subjective judgment with the objective nature of smart contracts. Although great for enforcing straightforward, measurable tasks, smart contracts struggle with subjective assessments of work quality. The question of how to fairly judge the nuances of Janet's work—like its efficiency or creativity—calls for a mix of technology and human insight. A possible fix? Adding a human-reviewed phase into the smart-contract process, where peers or managers assess the work. This combines the best of both worlds, ensuring fair and human-centered evaluations, especially in areas where the quality of work isn't just a numbers game.

Then there's the issue of complexity and skill set. Currently, most HR departments are not stocked with blockchain experts. A PwC survey indicated that only 5 percent of HR professionals have experience with blockchain technology.[10] Crafting a smart contract isn't like writing an email; it requires a nuanced understanding of programming languages. Additionally, cases like Janet's highlight the growing need for what some are calling "smart lawyers"—legal professionals trained to draft smart contracts that are not only efficient but also ethically and socially responsible. These experts must consider the multifaceted challenges and gray areas that exist in real-world employment scenarios, ensuring that smart contracts can adapt to the complexities of human life. These concerns pose a steep learning curve for companies looking to jump on the smart-contract bandwagon.

Fortunately, we aren't alone in our concerns. Organizations are already studying the maze of legal, technical, and ethical hurdles that come with this technology. Initiatives like the Enterprise Ethereum Alliance are in the early stages of creating standardized frameworks to ease some of these concerns by establishing clear guidelines and protocols that ensure the ethical use and legal compliance of smart contracts. This effort aims to harmonize the technology with existing legal structures and ethical standards, making it more reliable and trustworthy for employers and employees alike.

Despite these challenges, the allure of smart contracts in the employment sphere is hard to ignore. And we're still very much in the experimental stage—it's the Wild West of technological innovation. So, we ask: Are we truly prepared to swap out the nuance, flexibility, and emotional intelligence that human judgment offers for the rigid efficiency of automated code?

Entering the World of Smart Contracts

Employers are navigating an era when technology and data are becoming the backbone of business and workplace operations. More and more they are turning to technological solutions to securely collect, store, and analyze data, as well as to streamline transactions and payments, whether with other businesses, employees, or independent contractors.

Considering that smart contracts are still emerging, with new uses constantly being discovered, employers must carefully weigh the risks associated with their implementation and enforcement. As blockchain technology becomes increasingly integrated into our daily lives, it's likely that smart contracts will soon become commonplace, often operating unnoticed in the background. This implies that their adoption for employment practices is on the horizon, promising to reshape how we engage with work in fundamental ways.

Smart contracts represent the potential to upgrade the fabric of our work culture and social contracts. As we grapple with a rapidly evolving work environment—where remote jobs, gig roles, and AI are already reshaping norms—smart contracts become the glue. They have the power to automate everything from intricate payroll processes to benefits distribution, and in doing so, they could fundamentally transform our understanding and experience of work. This leads to a world where freelancers have the same security as full-time employees and where your employment terms can adapt in real time to employees' performance and needs.

The potential of smart contracts in revolutionizing the way we work promises to redefine workplace dynamics by automating and securing transactions with unmatched efficiency and transparency. This shift toward a more automated, trust-based system simplifies the complexities of contractual obligations and payments, which allows HR leaders to focus on innovation and growth. The journey ahead with smart contracts is not just about navigating challenges but embracing the vast opportunities they present for creating a more efficient and empowered workplace.

CONVERSATION STARTERS

1. Considering the application of smart contracts across various sectors, which use case resonates most with your organization's needs, and how might you pilot a smart-contract initiative to test its viability?
2. Where in your organization's workflow do you see the greatest potential for efficiency gains through smart-contract integration?

3. Considering the transformative potential of smart contracts for enhancing employee satisfaction, payment processes, and professional development, what specific initiatives could your organization undertake to harness these benefits?
4. Reflecting on the challenges and ethical considerations associated with smart contracts, what steps can your organization take to mitigate these risks while exploring smart-contract applications?
5. As your organization contemplates integrating smart contracts into its operations, what are the long-term strategic implications for your business model, employee relations, and competitive advantage in the market?

CHAPTER 8

Decentralized Work and the Next Iteration of the Gig Economy

Marcos Rezende was born in the countryside of Minas Gerais in Brazil and faced a series of personal challenges early in his life. His father died when he was eight, and he was later diagnosed with an autoimmune disease that required him to avoid sunlight. So, he turned indoors and discovered the world of computers and programming.

Fueled by his newfound passion, Rezende eventually became a web designer. He immigrated to Canada to further his education at the University of Alberta, which opened the door for his first user experience (UX) designer job at a top firm in Ottawa.

His career path was set before him, until life took another drastic turn when his two-year-old son was diagnosed with a rare form of cancer. Rezende's work-life balance was thrown into complete disarray. Having spent days and weeks at his son's hospital bedside, his full-time job proved inflexible and his employers were unsympathetic to his situation. Rezende faced a tough choice. Either he returned to work or faced termination at a time when his family needed him (and financial security) the most.

Looking for alternatives, Rezende discovered Upwork, a freelancing platform that changed everything for him. Here, he could take projects on his own terms, working from the hospital bedside of his ailing son or late into the night, providing flexibility to be there for his family. He was surprised at his earning power, making two-and-a-half times the earnings of his full-time job. Over time, Rezende became a leading figure in the UX community, contributing to well-known publications like *Career Foundry* and even earning a spot in the top 3 percent of global freelancers on Upwork. His website was nominated for prestigious awards, and his career soared—all while he maintained the work-life balance that was crucial for him and his family, especially as his son began to recover.[1]

Rezende summed it up best when he said, "I never would have unlocked my creativity in this way if I didn't leave corporate life for the bounty of freelancing."[2] And while a portion of today's workers have found stability in freelance work, it still isn't viable for the majority of today's skilled laborers because securing gigs demands significant effort, navigating stiff competition and often accepting lower pay for projects just to establish a solid client base.

The gig economy is poised for a transformation into a dynamic powerhouse, thanks to the advent of AI, decentralization, and skills-tracking platforms. These innovations promise to open up unprecedented opportunities for workers bold enough to embrace them.

From Gig Economy to Gig Sphere

Rezende's story exemplifies the life-changing flexibility and empowerment that the gig economy can offer, particularly in times of personal hardship. During our interview with him, he said, "One of the biggest challenges for me has been dealing with the hefty fees that these platforms charge. It eats into your earnings more than you'd expect. And then, there's this constant feeling of powerlessness when it comes to their policies—you just have no say. But perhaps what bothers me the most is how I don't truly get to own the work I pour my heart into. It's like, no matter how much you put in, you're always at a disadvantage, fighting to claim what you've rightfully earned."[3]

He's not alone in facing these challenges. High fees that can eat up to 20 percent of earnings, payment delays, and fierce global competition that drives down rates all present substantial hurdles for freelancers. Additionally, the lack of traditional employment benefits, such as health insurance and retirement plans, places a greater financial burden on freelancers.[4] The platforms' policies often leave freelancers with little control over their work, while the unpredictable flow of projects can result in unstable income. These issues underscore the need for freelance platforms to address the balance between offering opportunities and ensuring fair, sustainable working conditions for freelancers.

The transition to decentralized work platforms promises to remove these barriers by giving freelancers like Rezende a stake in their own professional ecosystem. Through the cutting-edge technologies of blockchain, smart contracts, and decentralized governance models, the platforms aim to elevate the gig economy to a more equitable and self-sustaining future.

More and more workers are shifting away from being replaceable cogs in the traditional corporate machine and are instead becoming unique, integral nodes in a decentralized ecosystem that some are calling the Gig Sphere. A study by Upwork revealed that more than $3.8 billion of work was done through Upwork by skilled professionals who are gaining more control by finding work they are passionate about and innovating their careers.[5] Additionally, the study showed that fifty-nine million Americans engaged in some form of freelance work in 2024.

To fully grasp the transformation that decentralized work platforms, we have to understand the limitations that plagued the traditional gig economy. Uber drivers and DoorDash delivery workers provide services we've come to rely on, but at what cost to them? The price is pretty steep. Companies like Uber and DoorDash operate as powerful middlemen, sometimes siphoning off as much as 30 percent of what the worker earns.[6]

Let's put that into context. Imagine you're an Uber driver who has just completed a day's work earning $200. Before you can even think about putting that money toward bills or savings, the platform whisks away $60. That's a significant chunk, especially considering that gig workers have to cover their own expenses, like car maintenance and fuel, taxes, and health care. According to an Economic Policy Institute report, gig workers earn roughly half the average wage of traditional full-time employees.[7] So, while it may seem as if gig workers have the freedom to "be their own boss," they often lack the financial stability that usually comes with it.

The hope of decentralized work platforms, by comparison, is to rewrite the employment equation in favor of the worker. What makes these platforms transformative is their ability to create a trustless environment. It means that the system itself ensures transparency and fairness, so you don't have to rely on a third party—like Uber or DoorDash—to act in your best interest. This ensures that hard-earned wages stay in the worker's pocket.

Every transaction, every agreement, every dispute is publicly recorded on the blockchain. This level of openness allows for a smoother dispute resolution process and deters any form of exploitation by either party. Imagine doing the same job as before but retaining 98 to 99 percent of your earnings instead of 70 percent. That's a well-deserved raise without increasing work effort or time.

The stats back this up as well. A study from the University of Cambridge found that workers on decentralized platforms reported feeling more empowered and financially stable compared with their counterparts on traditional platforms.[8] This is a game-changer in the realm of freelance and contract work, suggesting that the future of the gig economy may not rest in the hands of giant corporations but in decentralized networks that give power back to the people.

As we look at the evolution of work, it's apparent that decentralized platforms are setting a new standard, addressing many of the inequalities and limitations of the traditional gig economy. And for the millions engaged in freelance and contract work, this could mean a future where their work is valued more fairly and their financial stability is more of a reality.

Core Features of Decentralized Work Platforms

Just as Rezende experienced in his jump into the gig economy, peer-to-peer interactions will shift to trustless systems that ensure fairness and transparency. These foundational elements come together to form a radically different work environment. Let's unpack these features to grasp how they're revolutionizing the way we work.

Peer-to-Peer Interactions

In decentralized work platforms, the technology enables direct interactions between freelancers and clients, eliminating the need for a middleman—and the accompanying fees. The shift from a fee-laden, intermediary-driven model to a direct peer-to-peer framework offers palpable financial benefits to workers.

But it's not just about saving money. Direct interactions can also lead to higher levels of trust and long-term relationship-building between clients and freelancers. When a platform doesn't act as a constant intermediary, the relationship can evolve naturally, with the hope of leading to more repeat business.

However, this direct interaction model also has its drawbacks. While cutting out the middleman can lead to stronger client-freelancer relationships, it may also expose both parties to increased risk. The absence of a centralized platform means there's no built-in mechanism for dispute resolution, quality assurance, or payment protection. Freelancers may find it challenging to secure payment for completed work, while clients might receive subpar results with little recourse for corrections or refunds. Thus, while the prospect of direct interactions and trust building is appealing, it requires both parties to exercise greater due diligence and perhaps even develop new methods for mitigating risks and resolving disputes.

Despite the risks, this evolution of the gig economy offers a world where hard-earned money stays where it belongs, and professional relationships aren't curated by a third party interested in taking a cut.

Trustless Systems

As we discussed in the last chapter, "trust but verify" is now "just verify." Imagine a client that claims your submitted work isn't up to par just to avoid paying you. This becomes a "he said, she said" situation that can drag on for days, weeks, or even months. Now, how does the game change when it's "just verify"? In this new playing field, trust isn't something you earn—it's something you confirm. The project flow has been recorded and can't be altered or deleted. And that's not just a hypothetical benefit; according to a Deloitte report, one of the critical advantages of blockchain is the "enhanced security, trust, and integrity of transactions."[9] This kind of systemic trust is especially crucial for cross-border projects where legal jurisdictions can complicate disputes.

A freelance writer, for example, sets the terms of her agreement with a client and is coded into a smart contract. Once she submits her work and the client approves it, the smart contract self-executes, and the agreed payment is automatically released to her on completion. What if the client isn't happy with the work? Since all the terms and conditions are transparently recorded and perhaps even validated by community members or a decentralized oracle network—an entity that connects blockchains to external systems—there's no room for subjective interpretations of quality.[10] Both parties must clearly define and agree on the criteria for what is considered acceptable work before the contract is executed. When a project's success hinges on personal taste or aesthetic appreciation, the rigid, predefined criteria of a smart contract might not fully capture the nuances of client satisfaction or artistic intent.

The same goes for reviews. In traditional platforms, both parties might hesitate to give honest reviews for fear of retribution. But on a decentralized platform, where reviews are transparent and unalterable, the platform adds an extra layer of credibility and accountability to the system. So, instead of sinking into a quagmire of "Did they really mean what they said?" or "Can I trust this review?" the answers are all clearly there on the public ledger.

Tokenization and Staking Incentives

Anyone who has ever had to haggle over payment rates for a freelance gig can tell you how frustrating the process can be. The decentralized gig economy is redefining how we perceive and receive value for our work through tokenization and incentives. According to a study from the University of Cambridge, around 101 million people used cryptocurrency in 2023, a 189 percent increase from the year before.[11] Cryptocurrency certainly has had its ups and downs over the past decade, but it will gradually become a mainstream form of payment.

Tokens can be used in payment for completing tasks, but it's not a simple one-off. Many of the platforms offer what's called "staking" options. For example, if a graphic designer finishes a big project, the smart contract not only triggers immediate payment in the platform's specific cryptocurrency token but also allows the designer to stake their earnings in a communal pool. Over time, as the designer continues to stake, they could earn additional tokens based on the overall performance and growth of the community. This is revolutionary, as it encourages long-term commitment by granting both workers and employers a stake in the ongoing success of the platform.

Tokenization and staking mechanisms are disrupting traditional incentive structures. The old way of doing things—hourly rates, project fees, and so on—starts to feel a bit outdated. With tokens and staking, you evolve from trading time for money to investing in a platform that you believe in and, in return, it invests in you. Your value isn't measured only in hours clocked but in the quality and impact of your work, with financial incentives that have long-term benefits. This is the fascinating future of work incentives, where your career is less about just trying to make a living and more about making a valuable, lasting contribution on your own terms.

Flexible Work Arrangements

Unlike conventional job marketplaces where workers must apply for projects, compete with others, and then prove their worth to clients, decentralized platforms enable workers to simply open an app and begin working. Advanced AI technologies take over the task of matching workers with suitable projects and teams, all based on their verified skill sets. This modern approach to work allows individuals to tailor their work-life balance to a degree that was previously unattainable, offering a new level of personal freedom.

The impact of this newfound flexibility has the potential to drastically affect the global economy. According to McKinsey, the increased productivity and workforce participation that decentralized platforms are expected to usher in could add a jaw-dropping $2.7 trillion to the global GDP by 2028.[12] In essence, the advantages of decentralized work platforms will improve the work-life balance and financial situation for individual freelancers and create a ripple effect that could significantly boost economic output on a global scale.

That kind of worker input can produce more balanced, equitable, and effective governance rules. DAOs and the governance models they bring are, in a way, the ultimate expression of worker empowerment. They offer a sense of collective ownership and accountability that is unparalleled in traditional work environments. Leaders navigating this shift need to understand how to integrate the decentralized, democratic governance models into organizational structures, ensuring that the future we are outlining also aligns with the core values and goals of their organizations.

The Anything Economy of Talent

If YouTube ignited the creator economy, giving a platform to anyone with a camera and an idea, then decentralization and web3 technologies are setting the stage for an even more inclusive and diverse economic landscape. No longer restricted to video content, people can now monetize virtually any skill, hobby, or talent. Whether you're designing innovative car parts, selling physical products like toys, or even trading in virtual real estate, the "anything economy" opens up unparalleled avenues for income generation.

The promise of the anything economy has been here for a decade, and it's changing how we think about work, talent, and opportunity. Whether you're a business looking to tap into a diverse talent pool or a freelancer aiming to hone your skills on the global stage, emerging technologies are leveling the playing field. They are turning what used to be gigs into sustainable, secure forms of employment that anyone can access from anywhere. The future of work is no longer bound by geography. It's global and decentralized, and brimming with untapped potential.

Subscription models like Patreon have already made it easier for creators to build dedicated audiences willing to pay for exclusive content. Meanwhile, platforms like Odysee utilize blockchain technology to offer a more transparent and equitable payment system, for both creators and their audiences. This has the potential to revolutionize the revenue models, breaking away from the shackles of traditional employment to enable skilled laborers to work in ways that make sense for their lifestyle and take home a more substantial piece of the pie.

Perhaps the most transformative aspect of the anything economy is how it lowers the barriers to entry across various fields. For example, what if a truck driver had big dreams of becoming a fashion designer? In the traditional talent process, achieving this dream would require relocating to fashion capitals like New York or Los Angeles, enrolling in expensive design schools, and navigating a labyrinth of networking events, all while scraping by on minimum-wage jobs. Now, in the decentralized landscape of the anything economy, these barriers open to anyone who wants to design digital clothing and sell it as NFTs, build a reputation that might be more valuable than a traditional degree, and even crowdsource the financial support needed to pursue their passion. This democratization of opportunity redefines what's possible by offering a vision of a future where your career is limited only by your imagination.

The Global Talent Marketplace

In the past, the idea of connecting talent to work was a cumbersome, often manually intensive process. Job boards, recruiters, and networking events were the traditional avenues. But let's step into the present where web3 and AI have created agile, personalized talent marketplaces that are as dynamic as the people they serve. The platforms don't just match a résumé to a job description but leverage AI-driven insights to fit nuanced skills and career aspirations with real-world opportunities.

Take the example of platforms like Toptal that have already begun leveraging AI algorithms to create more dynamic matches between freelancers and businesses. Forget trying to find a Python developer and interviewing them to discover their preferences; now you can find a Python developer with experience in machine learning who's also interested in renewable energy projects and hire them within seconds. The specificity is groundbreaking. That's the level of granularity and personalization AI brings to the table.

Now, let's introduce another layer—personalized and dynamic development planning. The same AI algorithms that match talent to jobs can also guide professionals through personalized career paths. No more one-size-fits-all career advice. Each path is dynamically adjusted based on real-world performance and changing job market needs. LinkedIn, for instance, is moving beyond being a simple professional networking site to offer courses and tailored career advice based on its rich data insights. According to LinkedIn's 2024 Workplace Learning Report, 77 percent of learning and development professionals believe reskilling and upskilling is their organization's top priority.[13] What's better than having a tailor-made guide for each employee?

In a traditional setting, opportunities often resulted from whom you knew or, even worse, from office politics. But today's talent marketplaces provide what can best be described as "bilateral talent deployment," or more simply, a two-way street. Employees can express their career aspirations and areas of interest, while employers can quickly find internal candidates for new roles or projects, thereby democratizing opportunity within the organization.

What's the outcome of all these converging factors? Workforce agility. Businesses can pivot quickly because they have the right talent in the right roles. Employees are more engaged because they're doing work that aligns with their skills and ambitions. A study by the Boston Consulting Group found that companies that have applied "new ways of working"—which includes leveraging AI and talent marketplaces—saw a 40 percent increase in employee productivity.[14]

In summary, modern talent marketplaces are complex, adaptive ecosystems that are enabling the democratization of opportunity and introducing unprecedented agility into the workforce, all while aligning deeply with individual skills and career goals.

Emerging Decentralized Work Platforms

We've attended numerous metaverse and web3 conferences where the focus is often on case studies demonstrating the potential of emerging technologies. While it may still be too early to definitively validate the best use cases, waiting for perfect examples could mean missing the opportunity to innovate. Imagine asking Google to prove its worth while it was still perfecting its search engine. Despite this, there are companies making bold strides into this space and setting the stage for what's to come. Here, we'll explore some of these pioneering ventures.

Bounties Network

Bounties Network is a platform that connects freelancers with bounty programs for various tasks, all paid in ETH or tokens created using the Ethereum blockchain. The platform aims to create a fair environment for both bounty issuers and fulfillers, addressing the power imbalances that can sometimes occur in traditional freelance marketplaces. One of its standout features is its global reach, with community members located on all six major continents and speaking multiple languages.

The platform also stands out for its token-agnostic approach, allowing for payment in any Ethereum-based token. Bounties Network's mission is to democratize the freelance landscape by aligning the incentives of freelancers and employers. It's a welcome innovation, especially considering the World Bank's report that cross-border remittances hit $689 billion in 2018.[15] Traditional payment methods often incur significant fees and delays, but Bounties Network sidesteps these issues, making it truly global and inclusive.

The implications of Bounties Network's approach are far-reaching. As the gig economy continues to grow, similar platforms could become the norm rather than the exception. By providing a more equitable, democratic environment for freelance work, this platform could change the freelance landscape for the better, influencing future work arrangements worldwide.

Colony.io

Colony is a platform designed to facilitate the creation and management of DAOs that aims to make it easier for people around the world to build organizations together online. Colony offers features like cross-chain governance, allowing DeFi token governance to be supercharged by using it as your Colony's native token.

What's particularly intriguing is Colony's vision for the "future of the firm." It points out that traditional ideas about work, occupation, and even companies are undergoing a transformation. A decade ago, who would have thought that a taxi company with no drivers would be worth $90 billion, or that software nobody owns would generate millions every day? Colony posits that in the future, traditional companies may become obsolete, replaced by more flexible, decentralized structures.

Colony challenges our archaic perspectives on what a company should look like and how it should operate. As remote work and the gig economy continue to shape the future of employment, Colony offers a glimpse into an alternative future that could be more robust and adaptable to rapid changes. It's a revolutionary rethink of the concept of organized work itself. With the DAO sector showing remarkable growth, the questions around the future of work are intensely practical. Colony stands at this exciting intersection, offering solutions that could very well dictate the future norms of organizational governance and structure.

DeeLance

Leveraging web3 technologies, DeeLance aims to resolve many challenges that gig workers face, such as low pay and high platform fees. The blockchain element promises transparency and offers trustless systems where every review, work assignment, and payment can be verified on a public ledger. Add to that the power of tokenization, which goes beyond mere compensation by offering long-term staking incentives for quality work. DeeLance's DAOs are expected to manage assets worth billions, making this feature a compelling aspect of its governance model.

The cherry on top might be the platform's envisioned integration with the metaverse. While details remain sparse, the concept opens up possibilities like virtual coworking spaces or interactive client meetings, fundamentally redefining what remote work could entail. As a package, DeeLance presents itself as a shift in how we understand freelancing and, more broadly, the future of work. It's a compelling example that invites us to consider how emerging technologies can transform traditional work norms into something far more dynamic and equitable.

Challenges and Future Opportunities

As with any disruptive technology, recognizing the challenges is the first step in turning them into opportunities for even more profound innovations and reforms. Let's take a close look at the roadblocks that might prevent the evolution of decentralized work.

Decentralized work platforms introduce new complexities in regulatory issues, such as liability and worker rights, that current legal frameworks aren't fully equipped to handle. As the platforms gain popularity, there's an increasing need for new regulations or amendments to existing labor laws to clarify the ambiguities. Far from stifling innovation, effective regulation could actually boost the mainstream adoption of decentralized platforms by offering legal security for both workers and clients.

Scalability and Usability

Those who have attempted to transfer Bitcoin or other cryptocurrencies during peak usage times are often familiar with the slow transaction speeds that can occur. These scalability issues could significantly hamper the user experience, leading to delays in payments or contract executions. Such limitations could act as a bottleneck, slowing down the broader adoption of decentralized platforms for work opportunities.

Various solutions such as Layer 2 technologies, a network system that inherits the security of the blockchain it is built on top of, address these challenges.[16] They aim to enhance the speed and efficiency of blockchain transactions but are still in their developmental phases and are not yet foolproof. As these technologies continue to mature, they hold the promise of resolving the scalability issues that currently plague blockchain-based platforms.

Potential for Exploitation

While the absence of a centralized authority removes a single point of control, it also eliminates a centralized point of accountability. This raises critical questions about worker protection and oversight: Who is responsible for making sure workers aren't exploited or maintaining fair work conditions? The decentralization narrative necessitates the development of robust governance mechanisms that actively prevent any form of exploitation or abuse within the platform.

While the concept of community governance is empowering, allowing users to have a say in the rules and policies of the platform, it's not without challenges. There's a real risk of community governance turning into a "tyranny of the majority," where cliques or large groups can make decisions that adversely affect minority members. This is why it's crucial that governance models in decentralized platforms are designed with fairness, inclusion, and balanced representation in mind. Safeguards must ensure that the interests of all members, not just the majority, are accounted for in decision-making processes.

Moving toward the Next Era of the Gig Economy

The gig economy was indeed a catalyst in the revolutionary change in the world of work, but what's unfolding now is a comprehensive evolution toward decentralization to improve the system for all workers. This is less about replacing bosses with blockchain or trading office spaces for digital landscapes and more about fundamentally altering employee-employer relationships. Transparency evolves from a company policy to a coded feature, empowerment is a built-in mechanism, and opportunities are globally accessible.

Imagine a world where your work is not necessarily considered a job, but an asset; being a stakeholder in your capabilities changes how you interact with your work, how you value it, and how it values you. In this new framework, the meaning of job security could shift from long-term employment to long-term investable skills. It could mean that disengaged workers are finally able to be invested in their work because they have a direct stake in the ecosystem of their industry.

What makes these challenges exciting is that they unlock a more equitable, efficient, and empowered way of life. The way we work doesn't have to be the way it's always been. It can be better; it should be better.

CONVERSATION STARTERS

1. Considering the empowerment and financial stability reported by workers on decentralized platforms, how can your business model shift to incorporate these aspects and improve the gig worker experience?
2. What are the potential risks and challenges that might arise from cutting out the middleman in a decentralized work environment?
3. As the gig economy evolves into a more dynamic and equitable Gig Sphere, what innovative employment models can your organization explore to stay competitive and attractive to top talent?
4. What strategies should your organization implement to maintain quality assurance and dispute resolution in a decentralized work environment?
5. How is your organization planning to incorporate decentralized governance models or their principles into your company's governance structure?

CHAPTER 9

Ownership, IP, and Ethics in AI and Web3

In the bustling world of AI research, a scientist named Timnit Gebru has made some critical discoveries in the ethics of AI. Born and raised in Ethiopia, Gebru had always possessed a passion for mathematics and technology and earned a PhD in computer vision from Stanford University.

As Gebru delved deeper into the world of AI, she became increasingly aware of a glaring issue: bias and discrimination within AI algorithms. She realized that the algorithms, when trained on dogmatic data, could perpetuate existing societal inequalities. This realization ignited a fire within her, and she decided to dedicate her career to addressing the problem.

In 2016, while working at Microsoft Research, Gebru, along with her colleague Joy Buolamwini, coauthored a groundbreaking paper titled "Gender Shades," which exposed the gender and skin-tone bias in facial recognition technology, revealing that the technology performed poorly on women and people with darker skin tones.[1] The study garnered widespread attention and spurred discussions about the ethical implications of AI.

Gebru's commitment to ethics in AI only grew stronger. She cofounded the organization Black in AI, which aimed to increase the representation of Black researchers in the field. She also became a leading voice in the AI ethics community, advocating for transparency, fairness, and accountability in AI development.

One of Gebru's most notable accomplishments came during her tenure as a coleader of Google's Ethical AI team. In 2020, she and her team were working on a research paper that highlighted the environmental and societal impact of large-scale AI models.[2] The paper raised concerns about the carbon footprint of the models and their potential to exacerbate inequalities, but also raised controversy within Google by highlighting the risks of large language models, which are key to Google's business. Gebru stood her ground, refusing to compromise the integrity of her research, and published the paper, sparking a much-needed conversation about the ethical responsibilities of tech giants. Gebru was later fired from Google.

Her story serves as a stark reminder of the importance of addressing ethical concerns as we navigate the uncharted territory of rapidly advancing technology. In the rapidly evolving landscape of technology, the fusion of AI's cognitive capabilities with the decentralized, trust-enhancing properties of web3 promises unprecedented innovation. But this technological convergence also presents complex ethical and legal challenges that demand our careful consideration.

To navigate this landscape successfully, we must explore the ethical principles that guide responsible AI development, seeking to ensure fairness, transparency, and accountability in the age of automation. Simultaneously, the decentralized nature of web3 technologies like blockchain introduces novel challenges and opportunities around intellectual property (IP). We must examine how IP controls must adapt to safeguard creators' rights and incentivize innovation in a decentralized, collaborative, and rapidly evolving digital world.

A comprehensive understanding and discussion of these issues are paramount. This chapter aims to shed light on the emerging debates around ethics and IP, providing detailed insights, real-world examples, and practical solutions for each area of concern. The goal is to equip you with the knowledge and tools needed to navigate the challenges effectively, ensuring the responsible and equitable development and deployment of AI, web3, and other disruptive technologies.

Accountability and Transparency

In the ever-expanding realms of AI systems and web3 applications, accountability has emerged as a central ethical principle. Developers and organizations wielding these powerful technologies must shoulder responsibility for the outcomes they produce. This accountability begins with a commitment to transparent practices, a dedication to clear documentation, and a focus on explainable AI (XAI).[3]

In the realm of AI systems, transparency serves as the bedrock on which ethical development is built. It involves openly sharing the goals, methodologies, and data sources behind AI models and applications. This transparency extends not only to the AI community but also to the end users and stakeholders affected by these systems. By offering insights into the inner workings of AI, developers foster trust and invite scrutiny, which can help identify and rectify biases, errors, or unintended consequences.

Clear documentation is equally crucial in ensuring accountability. Developers must integrate online audit trails (such as those native to web3) of AI model development, data sources, and decision-making processes. The ledgers not only aid in understanding and debugging AI systems but also become indispensable when addressing ethical concerns or legal requirements. Online ledgers are critical reference points for accountability and continuous improvement.

XAI is a pivotal component in achieving accountability. It involves designing AI systems in a manner that makes their decisions understandable to humans. Transparent and interpretable AI models are essential for users and stakeholders to comprehend why an AI system makes a particular decision. In cases of disputes or adverse outcomes, explainability enables individuals to trace back to the root causes, enhancing accountability and fairness in AI deployment.

Here, smart contracts and decentralized governance mechanisms play a pivotal role in upholding transparency in decision-making. Since smart contracts execute predefined and event-driven agreements autonomously, they ensure that the logic and outcomes of the contracts are transparent and auditable, guaranteeing accountability in web3 applications.

Upholding transparency in the decision-making processes of decentralized organizations is also crucial in a web3 context to ensure that the interests of all stakeholders are considered and that power is not unduly concentrated. Accountability mechanisms within DAOs can include voting records, proposals, emails, direct messages, and public audits, which enable members to hold the organization accountable for its actions.

Accountability is a foundational ethical principle that must underpin the development and deployment of AI systems, web3 applications, and other advanced technologies. By prioritizing accountability, developers and organizations can navigate the complexities of these transformative technologies while fostering trust and ethical responsibility in the digital age.

Privacy and Data Security

23andMe, Marriott International, Equifax, Capital One, Target—few companies have been immune from cybersecurity attacks, which has driven users to question how AI and other technologies should handle and protect personal data to prevent such privacy breaches.

On centralized platforms, users often have a clear understanding of how their data is collected and utilized, but the decentralized nature of web3 challenges individuals to fully understand the implications of sharing their data across various blockchain-based applications. In the dynamic landscape of web3 technologies, where decentralization is a guiding principle, the concept of individual privacy takes on a new dimension. As the technologies enable decentralized data storage and transactions, they simultaneously raise significant ethical concerns regarding the protection of personal information.

Respecting the right to be forgotten, a fundamental tenet of data protection, also requires careful consideration within the web3 ecosystem.

Users should have the ability to erase their data from blockchain networks or other decentralized platforms when they no longer wish to be part of a particular system.

Some companies are making progress in this area. Decentralized platforms like Mastodon and Peepeth are embracing decentralization, giving users more control over their data and privacy by allowing them to own their content and decide how it's shared across the network. This approach represents a shift toward autonomy and consent at the forefront. By leveraging blockchain's inherent properties of transparency and immutability, the platforms are working toward creating a secure environment where personal data is protected, yet accessible under the right conditions. This is a delicate balance to strike, but it's crucial for the success of web3 in upholding the highest standards of privacy and data protection. In an evolving landscape, the conversation around data privacy is becoming more sophisticated, pushing the boundaries of what's possible in terms of secure, user-centric data management and sharing.

Bias and Fairness

In the world of AI, algorithms wield incredible power, shaping decisions in domains as diverse as finance, health care, and criminal justice. Yet, lurking beneath seemingly impartial calculations lies a profound challenge: the potential for bias. AI algorithms often mirror the biases embedded within the data they are trained on (and the ingrained system prompts that their owners order them to follow), inadvertently perpetuating discrimination and inequality. This conundrum presents us with an ethical imperative that cannot be ignored.

Consider that Amazon's AI recruiting tool, designed to help hire top talent, was scrapped in 2018. Why? Point-blank, the algorithm was biased against women. It favored male candidates because the AI tool was trained on résumés submitted over a ten-year period, which reflected historical gender imbalances in tech roles. The machine effectively inherited and amplified society's long-standing prejudices. Conversely, Google's high-profile Gemini launch in February 2024 was marred by *overcorrecting* for bias, delivering a mix of inaccurate and outrageous results by depicting historically inaccurate images of people, such as racially diverse Nazi-era German soldiers or a Black version of George Washington and the pope.

To combat this issue, developers must design inclusive data sets that accurately represent the diversity of the real world. It means carefully considering what data is included, how it is labeled, and whether any historical biases are unintentionally reinforced. To take it a step beyond, developers must also craft algorithms that not only learn from data but also adapt and self-correct to promote fairness. Imagine the enormity of this challenge.

Transparency and accountability are vital components of this ethical journey. To ensure fairness, AI systems and their developers must be open books, allowing external audits and scrutiny of the data used in a particular prompt. By leveraging the decentralized nature of web3, we can foster an environment where audits are not sporadic or conducted by a select few, but a continuous process accessible to a broader audience. This level of transparency ensures that biases, inaccuracies, or ethical lapses in AI systems can be identified and corrected promptly. Web3's inherent features offer a platform for recording and tracking the audits and changes in an immutable ledger to provide an objective and unalterable history of adjustments and improvements. This approach can enhance trust in AI technologies and empower users with clearer insight into how their data is being used and protected.

In essence, the mission is clear: AI should be a force for good, a tool that amplifies our humanity rather than exacerbates our flaws. Achieving fairness and equity calls for a concerted effort by developers, researchers, policy makers, and society at large to ensure that AI algorithms, like the societies they serve, are founded on principles of justice, equity, and compassion. This transparency not only builds a foundation of trust but also empowers individuals and communities affected by AI systems to seek redress when biases or discrimination occurs.

Governance + Autonomous Systems

The rise of autonomous features in vehicles, such as Tesla's Autopilot, has brought with it a host of questions that we can't ignore. After several accidents involving the AI-driven cars, some of which resulted in fatalities, the question of accountability has moved to the forefront of public discussion. Who should be held accountable when a machine, designed to make critical decisions, errs on the road? The incidents highlight the urgent need for establishing clear guidelines and frameworks that outline responsibility and governance in the realm of autonomous AI systems.

While the promise of autonomous vehicles holds tremendous potential for improving road safety and transforming transportation, these advancements clearly come with complex ethical dilemmas. As we continue to integrate AI into crucial aspects of our daily lives, we must develop robust ethical standards and accountability mechanisms. They should guide the technology's development and protect the public by ensuring that responsibility is clearly defined when autonomous systems are at the helm.

Job Security and Disruption

The transformative potential of AI and automation in the workforce has sparked unprecedented ethical questions, particularly concerning unemployment and social disruption. As machines become increasingly capable of performing tasks humans traditionally carry out, worry over job displacement and economic inequalities have taken center stage. Addressing these concerns requires a thoughtful approach that acknowledges the potential consequences of technological advancement.

One of the primary ethical responses to AI-driven job automation is a commitment to investing in workforce retraining and upskilling (something we'll discuss in more detail in chapter 11). As certain jobs become automated, it is essential to provide workers with the opportunity to acquire new skills that are relevant in the evolving job market. By offering accessible and high-quality retraining programs, leaders can help individuals transition into new roles, ensuring they remain employable and capable of contributing to the economy.

Developing policies to support displaced workers is another crucial response to AI-driven job displacement. Policy makers must design safety nets and social support systems that can cushion the impact of job loss. These policies may include unemployment benefits, healthcare coverage, and wage subsidies, among others. Such measures not only protect the livelihoods of affected individuals but also help mitigate the potential social disruption that can arise from widespread unemployment.

That's why the implementation of a universal basic income (UBI) is a bold and innovative ethical response to the challenges posed by AI automation. UBI entails providing all citizens with a regular, unconditional income, regardless of their employment status. While UBI has its proponents and critics, it has gained attention as a potential solution to ensure economic stability in an era of increased automation. Advocates argue that UBI can provide a financial safety net, allowing individuals to cover their basic needs and pursue opportunities for personal and professional development.

Via established tokenization plans, web3 offers incentive mechanisms to coordinate and align associated stakeholders in job retraining programs for today's AI era. International financial regulators' increased acceptance of web3 tokens also enables compensation for those aligned with AI retraining incentives to directly participate in UBI programs.

AI's capacity to automate jobs brings forth profound questions regarding unemployment and societal upheaval. As AI technology continues to reshape the job landscape, addressing these concerns will be crucial in promoting social equity and ensuring that the benefits of automation are broadly shared.

IP and Ownership

While AI opens up remarkable possibilities, it also makes it easier to infringe on existing IP rights, presenting an urgent need for updated regulations. One striking example came in 2021 when researchers developed an AI model capable of generating hyper-realistic images of Nike footwear. This technology was then exploited to produce counterfeit shoes that were virtually indistinguishable from authentic Nike products. In turn, AI is being used to identify counterfeits. It's a virtuous AI paradox.

Within the realm of AI, IP protection often revolves around safeguarding algorithms and models. Patents can cover novel techniques in AI, while trade secrets offer a way to protect proprietary algorithms. However, the tension between maintaining IP safeguards and fostering an open environment encourages innovation in AI. Striking the right balance between these two priorities is a growing challenge for both policy makers and technologists, as these technologies become more deeply embedded in our daily lives.

IP Ownership and Counterfeit

The decentralized architecture of web3 presents both challenges and opportunities when it comes to IP rights and creative works. Innovations like NFTs and decentralized content platforms have emerged, pushing the boundaries of how we think about copyright and ownership. For instance, since smart contracts can automatically manage royalty payments, they offer a new paradigm for fairly compensating artists. But in 2022, hackers replicated the Bored Ape Yacht Club, a billion-dollar NFT collection, and successfully sold counterfeit versions thanks to the decentralized nature of blockchain. Such a strike against an artist's original work highlights the challenges of enforcing IP rights in this new landscape.

The sale of Beeple's *Everydays: The First 5000 Days* for $69 million at a Christie's auction in March 2021 truly changed the game for digital art.[4] This artwork, a collage of five thousand individual images created daily over nearly fourteen years, was sold as an NFT, making it one of the most expensive pieces of digital art ever sold. The moment highlighted the power of blockchain technology to transform how we value and trade digital creations. This landmark case opened up a lot of discussions around the complexities of managing copyrights, ownership, and royalties within the NFT ecosystem. It was a stark reminder that as we navigate this exciting but uncharted territory, both the technology and the legal frameworks surrounding it will have to evolve to protect creators while encouraging innovation.

NFTs can teach us a lot about owning and protecting creative work in AI. Here's a simple way to look at it: Imagine if the data that helps AI learn could be linked to NFTs. The NFTs would ensure that every time AI uses someone's data, the original creator gets paid automatically. This could keep track of who owns what and make sure creators are fairly rewarded. It's like having a supersmart system that always remembers to say thank you with a payment every time it learns something new from someone's work.

AI-Generated Content

AI-generated artworks have surged in popularity, creating a murky terrain where traditional boundaries between creatorship, copyright, and ownership are unquestioningly blurred. Discussions are underway about how current IP laws should evolve to account for these novel forms of creation, but there's no consensus yet on how to adapt existing frameworks to the emerging realities.

One landmark case that spotlighted the challenges involved a group of artists and AI researchers called Obvious, who in 2018 created a portrait using deep learning algorithms.[5] Titled *Portrait of Edmond de Belamy*, the artwork left open the question, who actually owns it—the AI, the human artists, or some hybrid entity? Then, in 2022, an AI-generated painting sold at auction for $432,000, yet Robbie Barrat, the artist responsible for creating the AI that produced the painting, received no royalties. These cases underscore the growing need for IP laws to catch up with technological advancements, as who holds the copyright for such AI-generated content remains ambiguous.

The essence of this dilemma lies in AI's ability to amalgamate and transform preexisting works into something novel, often without explicit acknowledgment or compensation to the original creators. This reality pushes the boundaries of fair use and raises concerns about the uncredited utilization of human-generated works.

Then there is the loss of IP. What if gen AI is building on someone else's existing artwork, writing, or photography? And how is that person's IP being infringed on or going uncredited? OpenAI was hit with a class action copyright lawsuit claiming its enormously popular AI chatbot ChatGPT was trained on 183,000 books, without permission from the authors.[6] When asked about it on the *Late Show with Stephen Colbert*, famed astrologist Neil Degrasse Tyson, who was told his fifteen books were used in its training, said, "I was honored by that, but then a little creeped out about it. Why do I write books? So you can read the books and do stuff with what you learned from the books. So now, if you can learn stuff quicker by having AI grab it, I . . . still have to think about that."[7]

This scenario necessitates a reevaluation of copyright laws to ensure they encompass the nuances of AI involvement in creative processes. It prompts a call for innovative legal solutions that recognize the contributions of both human creators and the technological tools that extend their creative reach to make certain fair compensation and credit are maintained in the fast-evolving landscape of AI-generated content.

As we navigate the confluence of human ingenuity and machine intelligence, we need to facilitate a symbiotic evolution to establish mutual growth and understanding between humanity and artificial constructs.

Patents

In 2021, IBM made waves in the IP community by filing a patent application for a new food container. What set this case apart was that the inventor listed wasn't a human, but an AI system called a device for the autonomous bootstrapping of unified sentience (DABUS). Created by Steven Thaler, DABUS is essentially a sophisticated collection of code and algorithms. This unconventional application sparked a debate on the legal recognition of AI as an inventor, forcing the US Patent and Trademark Office (PTO) to confront uncharted territory in patent law.[8] Thaler had initially filed two patent applications in 2019, both listing DABUS as the sole inventor and stating that the invention was "generated by artificial intelligence." However, the PTO rejected both applications, insisting that a valid inventor must be a human.

Thaler challenged the PTO's decision, initially through internal reviews and eventually in district court. This led to a protracted legal battle that culminated in a landmark decision. The central question revolved around whether the term "inventor" under US patent law could be extended to nonhuman entities. The US Patent Act defines an "inventor" as "the individual . . . who invented or discovered the subject matter of the invention," but it doesn't specify what constitutes an "individual." The matter was finally settled on August 5, 2022, when the US Court of Appeals for the Federal Circuit definitively ruled in *Thaler v. Vidal* that AI systems could not legally be considered inventors for US patents, setting a precedent that reinforced the anthropocentric view of IP rights.

As we venture further into an era where AI's capabilities continuously blur the lines between human- and machine-made creations, we need to reconsider our long-standing principles surrounding innovation and ownership. The ruling underscores an essential legal principle: patents are reserved as incentives for human ingenuity by reflecting a societal value in recognizing and rewarding human creativity and labor. Yet, it also highlights a significant ethical consideration; as AI systems like DABUS become more sophisticated, contributing creatively in ways previously thought exclusive to humans, how do we ethically attribute credit and ownership? This presents an urgent call for policy makers, legal experts, and technologists alike to collaboratively forge new frameworks that can accommodate these advancements.

Open Source and Collaboration

The open source ethos is a cornerstone of many AI and web3 projects, as it nurtures collaboration and rapid innovation; however, this approach poses a complex set of IP challenges. Developers need to tread carefully through a labyrinth of licensing agreements, ensuring that they give due attribution while also safeguarding the integrity of their work. Open source doesn't mean free-for-all, and understanding the nuances of various licenses becomes essential for both contributors and users.

The web3 community was born out of open source and remains deeply rooted in it. Most of the most popular web3 projects (such as the audited and trusted OpenZeppelin NFT smart contracts) are fully open.[9] The transparency, accountability, and "build in public" community contributions to web3 software are endemic to the industry, representing the best of the open source ethos.

Within the realm of AI, the IP conundrum extends deeply into the algorithms and models that power these technologies. Companies and academic researchers pour substantial resources into creating novel AI solutions, and naturally they seek to protect their investments. Take, for example, OpenAI's GPT, a language model that has set new benchmarks in the field. Though it sprang from a culture that values open sharing of information, GPT is proprietary technology. OpenAI maintains strict control over its use, offering licenses for specific applications, which underscores the delicate balance between open collaboration and IP rights in the evolving landscape of AI and web3.

Tokenization of Intellectual Assets

Tokenization, another aspect of web3, allows for the fractional ownership and trading of assets. This concept extends to IP, where patents, trademarks, and copyrights could be tokenized, enabling broader access and investment opportunities.

For instance, companies like Maecenas use blockchain technology to tokenize fine art, enabling multiple investors to own fractions of valuable artworks. IPwe is a platform that aims to tokenize patents, making them more accessible to a global audience of investors. Mattereum is working on tokenizing physical and intellectual property, aiming to create a bridge between the physical and digital worlds by using blockchain and smart contracts to enforce the IP. Some DAOs, which are a hallmark of web3, explore tokenized governance and IP ownership. DAOs can collectively own and manage IP assets, enabling decentralized decision-making regarding their use.

Tokenizing IP has the potential to democratize access, increase liquidity, and transform the way IP assets are managed and monetized. However, it also raises legal and regulatory challenges that need to be addressed, as this innovative approach to IP ownership and management continues to evolve, like determining how to enforce IP rights when assets are tokenized and distributed across a global blockchain network and navigating the complex web of international copyright laws that may apply to these digital assets. Standard regulatory frameworks are still in development, which leaves unanswered questions about taxation, securities law compliance, and consumer protection.

Data Ownership and Personal Information

There are novel challenges in the realm of IP, particularly when it comes to data ownership. Take, for instance, a decentralized social media platform where users are continually posting content. In a centralized world, the platform generally owns—or at least controls—the data. However, in a decentralized setting, data control can often be distributed among the community of users. This disrupts traditional frameworks of data ownership and IP rights, adding layers of complexity as data becomes more decentralized and community-governed.

This emerging landscape demands a rethinking of data ownership models, especially when it comes to personal information and identity. Web3 technologies are revolutionizing the way we manage our digital identities by reducing our dependence on centralized platforms. Solutions like self-sovereign identity, exemplified by blockchain-based platforms such as Sovrin and uPort, are giving individuals unprecedented control over their personal information. These technologies enable users to manage their digital identities securely, providing a safeguard against identity theft and unauthorized data access.

However, the transition to decentralized identity management isn't without ethical problems. The rise in cases of identity theft and frequent data breaches underscore the critical need for robust, secure solutions for digital identity management. As these web3 technologies gain traction, it's essential to address challenges related to identity security and ethical usage to ensure that this shift genuinely empowers individuals without compromising their safety.

Protecting AI Algorithms

The rapid evolution of disruptive and decentralized technologies necessitates a careful balance of ethical, legal, and IP considerations. Developers, organizations, policy makers, and society at large must collaborate to ensure responsible development, equitable access, and protection of propriety and personal data and intellectual creations. Striking this balance will be critical in realizing the full potential of these transformative technologies while safeguarding against their potential harms.

The bottom line here is that the implications within AI and web3 technologies are multifaceted and continually evolving. While the technologies open up new frontiers for innovation and collaboration, they also challenge traditional norms and call for adaptive legal and ethical frameworks. Addressing these challenges will be essential in fostering responsible innovation and equitable access to intellectual creations in the digital age.

CONVERSATION STARTERS

1. Given the complex legal landscape surrounding AI-generated content and the potential for copyright infringement, how is your organization navigating these waters to respect original creators' rights and ensure fair compensation?

2. How can your organization contribute to developing ethical AI applications that minimize bias and ensure fairness across all demographics?
3. How is your organization leveraging web3 technologies to empower users while safeguarding their personal information against misuse and breaches?
4. As the boundaries of IP expand with AI and web3 innovations, how is your organization staying abreast of legal developments to protect its interests and respect others' rights?
5. What vision does your organization hold for the future of AI and web3 technologies, and how is it actively working toward realizing that vision in a responsible and equitable manner?

PART IV

Our Evolving Relationship with Work

CHAPTER 10

Building an Inclusive Work Culture for a Decentralized Workforce

A Year Without Pants. That's the title of Scott Berkun's book detailing his experience as a manager at Automattic, the power behind WordPress.com, which powers an astonishing 835 million websites, or more than 40 percent of all active sites globally. It's not that he doesn't choose to wear pants, although that's his prerogative; it's that he can work from anywhere in the world he wishes, barely uses email, and his team launches improvements to its products dozens of times a day. With a fraction of the resources of Google, Amazon, or Facebook, it has a similar impact on the future of the internet. Automattic mastered the art of remote work long before it became a global necessity. With employees sprinkled across more than seventy countries, Automattic has turned geographical diversity into its greatest asset in forming a work culture that's as innovative as it is inclusive.

From the get-go, it embraced the philosophy that work isn't about where you are, but what you do. This ethos allowed it to harness the power of asynchronous communication to prove that productivity isn't bound by time zones or office walls. It armed its team with the tools and resources needed for a seamless home office setup—from technology stipends, to access, to cutting-edge software—and empowering each member to create their ideal work environment.

What is Automattic's secret sauce? A transparent culture where information flows freely, making every remote worker feel connected to the company's heartbeat. It has regular team meetups and adventures spanning the globe from bustling cities to tranquil retreats, creating bonds and turning colleagues into a tight-knit community. And through it all, it has a steadfast commitment to well-being ensuring that every employee feels supported, both professionally and personally.

As the world finds its postpandemic rhythm, many companies have chosen to return to a physical office, placing the highly cherished benefits of remote work in a precarious situation. Despite the widespread adoption of hybrid models, successfully merging the benefits of in-office and remote work, there's an ongoing debate about maintaining the essence of remote work's advantages. A striking finding from a 2024 Gallup report reveals that an overwhelming 94 percent of employees favor a hybrid work setup, which underscores the need for distributed work arrangements.[1] Leaders need to think carefully about a more distributed and flexible workforce that is integrated into the evolving work culture.

Recent statistics from Live Data Technologies show the challenges faced by fully remote workers, who were promoted 31 percent less frequently than their counterparts who were part-time in the office in 2024.[2] This disparity points to a broader issue within digital work environments: remote employees often experience fewer meaningful work connections, lower pay, and fewer opportunities for advancement compared to in-office workers. These challenges highlight a prevalent mindset among leaders, many of whom either overlook the benefits of a distributed workforce or struggle to engage remote employees. Decentralized work amplifies these concerns, as employees immersed in such environments experience a sense of disconnection, which emphasizes the urgent need for strategies that bridge the gap and encourage a sense of belonging and equity, regardless of where work is done.

Creating an inclusive work culture in today's digital era demands that we fundamentally rethink our approaches to leadership, how we assess performance, and the ways we support our employees. To begin, let's examine the essential human needs that employees are increasingly calling attention to.

The Human Deal

In the digital age, the traditional employee value proposition is transforming. Gone are the days when a paycheck and a corner office were enough to secure loyalty and engagement. Employees want meaningful work, a sense of community, a balanced life, and recognition as humans at work, not just resources. This shift is so significant that Gartner termed it "The Human Deal," and it's changing the rules of employment in the digital era.[3]

Employees are seeking radical flexibility that allows them to work when and where they're most productive. Although digital tools empower us to work from anywhere, they should also incorporate all the elements that make us human at work, full of rich human interactions, serendipitous connections that spark creativity, and a sense of purpose that unifies the community.

The opportunity is to embrace a more human-centric approach that values employees for who they are, not just what they can do. This involves rethinking compensation structures and in-office requirements, but it also means creating a culture of trust and empowerment. It means allowing employees to bring their authentic selves to work—however and wherever they do so—investing as much in their holistic well-being as their professional skills.

The prevailing work model, focused on relentless productivity, has severely limited the opportunities for meaningful work, especially in Western societies. This dynamic often feels one-sided, as if the system is extracting value from workers while neglecting the human qualities that make work fulfilling in the first place. It seems to lack the elements that make us fundamentally human. After all, we only have one life. Why are we willing to squander it doing unfulfilling work? Shouldn't the quest for value and meaning extend beyond the boardroom to touch the lives of those who contribute to that value every single day?

In his book *Love + Work,* Markus Buckingham shares an important lesson: "At work, according to the most recent data, less than 16 percent of us are fully engaged, with the rest of us just selling our time and our talent and getting compensated for our trouble."[4]

As our lives continue to digitize, the conservative boundaries that once defined the workplace are dissolving. And when you combine these challenges with the rise of emerging technologies discussed throughout this book, it's easy to feel as though the workplace is becoming increasingly devoid of human connection and meaning.

However, we argue that this perception is far from reality. The whole point of the decentralization movement is aimed at reintegrating the human elements that often seem lost in today's work environments. It's also about recognizing the unique qualities that make us human and cultivating a work environment that respects and amplifies those qualities.

These emerging technologies offer us a unique opportunity to create workspaces that are more human-centric and meet our needs and aspirations in ways we might not have imagined. Embracing The Human Deal prompts organizations to transform how they engage with their workforce, fostering a culture where individuals flourish, no matter where they're based. With this foundation, let's explore three critical pillars essential for nurturing an inclusive distributed work culture: connection, work environment, and well-being.

Connection

The great paradox of a distributed workforce lies in the physical distance that separates us, challenging traditional notions of connection, yet simultaneously offering new avenues for human and meaningful interactions through emerging technologies. Despite being miles apart, the innovations bridge the gap that makes communication and collaboration feel as real and engaging as in-person encounters. While we may be physically distant, technology empowers us to maintain, and even enhance, our connections to preserve the essence of teamwork and camaraderie that thrives in the digital age.

From Remote Work to Full Distribution

At first glance, it might seem that transitioning from a traditional office setting to remote work is the final destination for modern work environments. However, as we've discussed in earlier chapters, we are already witnessing the advent of something far more radical: fully distributed teams working within a DAO. In contrast to remote work, where the office still exists but employees can work from other locations, fully distributed teams have no central office at all, and no central decision-maker.

Consider that Buffer, a fully distributed company, found in its "State of Remote Work" report that 98 percent of respondents would like to work remotely, at least part-time, for their entire careers.[5] The report reveals that the top benefits people see in the opportunity to work remotely are a flexible schedule (32 percent), the ability to work from any location (26 percent), and the lack of a commute (21 percent).

But full distribution offers even more: it democratizes opportunity. Talent is everywhere; opportunities should be too. When a company is fully distributed, it can tap into a global talent pool, unbounded by geography or the need for relocation. GitLab, for example, has embraced a fully distributed model and has team members in more than sixty-five countries.

In fully distributed work models, we see the ultimate realization of digital work's promise: to free us from the constraints of time and place so that the focus becomes purely on what we produce and how we collaborate. Indeed, in a fully distributed environment, work is something you do, not a place you go.

Collaboration in the Metaverse

As the concept of the metaverse ebbs and flows in the mainstream media, the idea of a collective, virtual, shared space, created by the convergence of virtually enhanced physical reality and interactive digital spaces, is still highly intriguing to workers across the globe. Various companies are experimenting with metaverse platforms to hold meetings, brainstorm, and even host corporate events. According to a report from PwC, virtual reality and augmented reality could add as much as $1.79 trillion to the global economy by 2030, much of which will be through improved modes of collaboration and work-related activities.[6]

The benefit here is the improvement in well-being and efficiency that comes from a more engaging and less monotonous work environment. A study by Stanford University found that the creative output increased by an average of 37 percent when participants were walking as opposed to sitting.[7] And while we admit it's not the same, the metaverse offers countless such opportunities to simulate environments that can boost creativity and well-being simultaneously. Imagine, for instance, brainstorming sessions held on virtual beaches, team meetings atop simulated mountain peaks, or individual focus time in serene, virtual gardens. Virtual offices like SoWork and Virbella are expanding these possibilities.

The metaverse allows for a form of social interaction that's been severely limited in the age of remote work. The metaverse could replicate, and even enhance, the social aspects of an office, like a virtual Ping-Pong table or a digital kitchen where you can sit with colleagues. Team members with disabilities, who might find traditional office settings limiting, can engage more freely in a virtual environment tailored to their needs. Meta, for example, showcased its Horizon Workrooms through the lens of a man with no legs who could roam fearlessly throughout its virtual office space and engage with coworkers in ways he couldn't in the physical world.[8]

So, what we are looking at is the future of collaborative well-being. The metaverse provides an avenue to enhance the human experience of work, redefining teamwork in a way that is holistically beneficial to everyone involved.

Virtual Team Building: Creating Cohesion Online

Virtual team building has become an art form that requires intention and a deep understanding of the team's culture. The pivot to remote work has challenged managers to rethink how to foster team cohesion in a digital landscape. Innovative solutions such as virtual escape rooms and online team challenges have proven effective in breaking down barriers and building camaraderie among team members whose locations may span different time zones. According to a study in *Harvard Business Review*, virtual teams can outperform their in-office counterparts when they are managed with a focus on building strong, interpersonal connections, underscoring the potential of well-curated online team-building activities to enhance productivity and job satisfaction.[9]

Companies like Zapier and Buffer have set benchmarks in virtual team building by integrating regular virtual coffee breaks, where team members can gather informally to discuss non-work-related topics, and remote retreats, in which team members meet in different parts of the world to work and bond. These practices help not only maintain a sense of belonging and community but also foster a culture of openness and trust. A report from *MIT Sloan Management Review* highlights the importance of creating an inclusive culture that supports virtual connections, suggesting that teams with a strong sense of trust and inclusion exhibit higher levels of innovation and engagement.[10] Thus, by adopting and adapting virtual team-building strategies, organizations can create a cohesive and vibrant work environment that transcends physical boundaries.

Serendipitous Interactions: Engineering Chance in a Structured World

Bringing back the proverbial watercooler chat could really boost collaboration. Serendipitous interactions, those chance encounters that fuel creativity and innovation, have proven to lead to greater connection and creativity, and arguably, they are one of the biggest losses of going remote.

Some forward-thinking companies are engineering these moments of serendipity with intention and ingenuity. Google, for example, famously designed its office spaces to maximize "casual collisions" among staff, a strategy based on the belief that the most innovative ideas often emerge from random, unplanned interactions. Now it is building that same consideration into its Azure platform to boost virtual serendipity.[11] Research supports this concept; a study published in *American Sociological Review* found that workers who engaged in more frequent, informal communications were more likely to generate innovative ideas, highlighting the tangible benefits of designing workspaces that encourage spontaneous encounters.[12]

In the digital realm, platforms like Donut, an app integrated within Slack, randomly pairs employees for virtual coffee chats, mimicking watercooler conversations. These digital tools bridge the gap between remote team members to gain a sense of community and belonging that is critical for collaborative creativity. By intentionally creating spaces and opportunities for serendipitous encounters, leaders can cultivate a culture where innovation flourishes, even in the most structured environments, proving there's always room for the unexpected to spark brilliance.

Now that we've explored connections in the future of work, let's dig into new work opportunities.

New Work Opportunities

Digital platforms expand the boundaries of collaboration and creativity and open the door to innovative new work opportunities—from global freelancing to roles that exclusively exist within virtual environments. Adopting an ambitious approach to integrating cutting-edge technologies can help you shape a future where work is more adaptable and aligned with the diverse aspirations and skills of a global workforce.

Digital Platforms That Support the Future of Work

Remember Discord, the breakout star in the evolution of the digital workplace? Initially gaining traction as a free voice, video, and text chat platform, Discord has quickly amassed an impressive user base of 150 million people logging in monthly. While some originally branded it a "GenZ Slack," Discord is morphing into something far more transformative and useful in a work context. It is the digital infrastructure for a new kind of organizational model known as a crypto network, or a group that forms around crypto protocols to enable new ways of coordinating, measuring, and rewarding contributions to complex ecosystems. Crypto networks create better alignment between participants, and DAOs serve as the coordination layer.[13]

Discord is being leveraged as a hub where people can gather not just for casual conversations but also for serious work and decentralized governance, particularly in the realms of blockchain and cryptocurrency. The very concept of a traditional, top-down organizational structure is challenged by the crypto network model, where power, decision-making, and even revenue can be distributed more equitably among participants. Discord's adaptability and user-friendly interface make it an ideal platform for these kinds of forward-thinking, decentralized communities.

As we look toward the future of work, platforms like Discord offer a new way to communicate that is reshaping our very understanding of what an organization can be. This horizon is about fostering environments where openness, community-driven innovation, and shared success are the norms, not the exceptions.

Digital Nomadism: Embracing Flexibility

The old adage "Location, location, location" has long been the mantra of real estate and even job markets. But that tune is changing to something more like "freedom, flexibility, connection." Welcome to the era of digital nomadism, a phenomenon reshaped and supercharged by the advent of decentralized work models.

Imagine waking up in a beachside bungalow in Bali, spending the morning working on a blockchain project for a startup in Sweden, then flying to Mumbai to connect with a new client—all while being a citizen of Canada. This is the reality for an increasing number of professionals. According to a study by MBO Partners, the number of digital nomads in the United States alone grew from 7.3 million in 2019 to an estimated 17.3 million in 2024.[14]

Beyond the individual lives and careers that are being transformed, the mass mobilization of remote work is disrupting traditional work models in multiple sectors, most notably real estate, city planning, and even national immigration policies. Cities like Dubai and countries like Barbados have already rolled out digital nomad visas, aiming to attract this new class of global mobile workers. This shift diversifies their economies and introduces fresh perspectives into local communities. For the first time, nations are essentially competing for citizens, a huge departure from how we've traditionally understood national immigration policies.

In the real estate sector, the rise of digital nomadism is upending traditional demand for office spaces, leading to an increase in coworking spaces and a reconsideration of long-term leases. City planners are taking note, too, contemplating how to develop infrastructure that supports transient populations, from more public Wi-Fi zones to short-term housing solutions.

What does this all mean for the adventurous emerging workforce? Essentially, as the technologies continue to untether work from geographical constraints, we are entering a new chapter in human history where freedom, flexibility, and meritocracy become table stakes. People are no longer confined to working in cubicles or even major metropolitan areas; they can truly work from anywhere, affecting everything from our daily commute to national GDPs. The potential for innovation and change is unprecedented, thanks in large part to a global marketplace that's becoming more inclusive and more accessible than before.

From Office Cubicles to Virtual Reality Workspaces

The traditional office cubicle has been a staple of corporate life for decades—beige walls and all. It's the setting many of us associate with the nine-to-five grind. Yet, technology—particularly precursors to the metaverse—has been staging a quiet revolution, transforming how and where we work. The jump from a physical office to an AR/VR workspace has yet to cross the chasm of adoption in our working environment, despite its profound implications for efficiency, collaboration, and overall well-being.

Now integrate VR capabilities, and suddenly you're not limited to a Zoom rectangle but can move and interact in three-dimensional digital spaces. You could be brainstorming on a virtual whiteboard one minute, and next find yourself in a simulated lab conducting experiments alongside colleagues situated on different continents. These are not futuristic scenarios; Apple and Meta are already providing such solutions within the immersive internet.

There are concrete benefits. A PwC study found that employees trained using VR were 275 percent more confident to act on what they learned after training, compared with classroom learners.[15] The level of immersion that VR provides can improve focus, understanding, and application of new knowledge. It opens up new avenues for workforce training, talent development, and even reskilling—important factors when the half-life of skills is shrinking.

And in the global marketplace, the decentralized office becomes truly decentralized when you can work efficiently and collaboratively from anywhere—your home, a crowded coffee shop, or a destination resort. This has huge inferences for everything from urban planning to immigration policies. Cities previously considered "non-hubs" could see a surge in populations and investment. A study from Upwork revealed that as many as fourteen to twenty-three million Americans are planning to relocate due to remote work opportunities.[16]

The metamorphosis from office cubicles to VR workspaces is an exciting evolution that challenges the very core of how we define employment. Enabled by gen AI and fortified by web3 technologies, we're on the brink of a work renaissance that promises to be more equitable and globally connected.

Innovative Employment Models: Beyond the Nine-to-Five

The traditional nine-to-five workday has been evolving beyond its rigid constraints as businesses and employees demand more flexibility and autonomy. Flextime arrangements, for instance, are being reimagined within decentralized frameworks. Companies like Patagonia are at the forefront of leveraging flexible models to enhance employee satisfaction and retention, illustrating the potential of a work culture unbound by time constraints.[17]

Chronoworking is an emerging trend that enables employees to tailor their work schedules according to their individual circadian rhythms rather than adhering to the traditional nine-to-five or eight-to-five workday. "This concept prioritizes recognizing and respecting your body's natural cycles of energy highs and lows, allowing for a more flexible schedule that accommodates peak periods of performance," according to corporate wellness consultant Tawn Williams.[18] "By aligning work tasks with these optimal times, individuals can foster a harmonious synthesis of mental, physical and emotional well-being, leading to enhanced productivity and overall job satisfaction."

The four-day workweek concept is gaining momentum; Microsoft Japan's trial revealed a 40 percent jump in productivity, hinting at the untapped potential of condensed work periods enhanced by focused, metaverse-enabled collaboration sessions.[19] Such models demonstrate that with less time constrained by traditional work setups, employees can achieve greater efficiency and innovation.

As we explore decentralized work models and embrace the metaverse for daily tasks, companies are discovering the extensive benefits the technologies bring, including heightened productivity, improved well-being, and robust talent retention.

Well-Being

The well-being of distributed employees, or decentralized contributors, as they are called in a web3 context, is a multifaceted challenge that encompasses mental, emotional, and social dimensions. Contributors report feelings of isolation, blurred work-life boundaries, and increased stress due to constant digital connectivity. To address these issues, companies are adopting holistic wellness programs that include access to mental health resources, initiatives to foster social connections, and policies that reach beyond geographic boundaries. By prioritizing the well-being of remote workers, organizations can build a resilient and engaged workforce that feels supported regardless of the physical location.

Building an Inclusive Company Culture for Remote Workers

The challenge of preserving employee well-being in remote or hybrid work settings is becoming increasingly prominent and has the attention of HR departments and business leaders. How can we sustain a robust company culture when team members are scattered across miles or continents?

Companies that recognize the scale of this issue are creating solutions to balance productivity and employee well-being. Some have implemented "No Meeting Fridays" to allow employees dedicated time for focused work, professional development, or simply a break from the screen. Others are leveraging AI-driven analytics tools to monitor employee engagement and identify potential burnout risks before they escalate. Firms are using AI to analyze employee surveys, feedback, and even Slack conversations to gauge employee sentiment and overall well-being.

The role of HR departments has expanded to include being mental health and well-being advocates. In addition to organizing regular check-ins and offering mental health resources, many are partnering with telehealth providers to offer virtual counseling services. According to a 2024 report by the National Business Group on Health, 77 percent of surveyed employers plan to implement virtual mental health services, representing a noteworthy increase from previous years.[20]

Businesses are in the midst of seeking that delicate balance between operational efficiency and the mental, emotional, and physical health of their employees, and HR leaders have been at the forefront of this change by stepping into the unknown to implement innovative approaches to foster a work environment that is not only productive but also humane.

Companies like GitLab have gone entirely remote and offer "Family and Friends Days," when the entire company takes a day off, preventing the guilt an employee might feel if only they took a personal day.[21] While these solutions aim to make remote work more sustainable, they also require a cultural shift in how companies view productivity and success. Rather than focusing solely on output, they emphasize outcomes and employee well-being as key performance indicators.

Physical Well-Being for Remote Workers

For many of us, the transition to remote work has blurred the boundaries between work and home life, making it more important to consider the physical setup of our home offices. An article in *Harvard Business Review* suggests that an optimal home office setup that mimics the ergonomics of a well-designed office space can significantly reduce the risk of physical strain and injury.[22] The setup includes finding the right furniture and placing monitors at eye level, the use of external keyboards to maintain proper wrist alignment, and the importance of good lighting to prevent eyestrain.

Moreover, digital well-being tools and apps, like Eye Care 20 20 20, which reminds users to take breaks and look away from their screens to prevent eye fatigue, are becoming indispensable in digital work routines. Encouraging employees to integrate such tools into their daily work can have a profound impact on reducing screen-time fatigue and promoting physical health.

Beyond the basics of ergonomic setup and digital health tools, companies can further support their remote employees in innovative ways. Transitioning from traditional perks like company cars to more relevant benefits such as standing desks, walking treadmills, or resources for becoming a digital nomad reflects an understanding of the unique needs of remote workers. With the rise of digital worker visas in countries like Japan and South Korea, companies have a new avenue to support employees wishing to explore work abroad, thereby enriching their work and life experience.

Creating work environments that prioritize physical well-being fosters a more engaged and productive workforce. By advising about optimal home office setups, integrating wellness technologies, and reimagining physical workspaces to encourage movement and relaxation, employers can make significant strides in promoting the overall health of their employees.

The Digital Detox: Balancing Screen Time and Productivity

In our digital-first environment, our lives are intertwined with screens. From the moment we wake up to the minute we go to bed, we're bombarded with emails, messages, notifications, endless streams of content, and the disruptive power of blue light emitted from screens. This constant connectivity can take a toll on our mental health and productivity, making the concept of a digital detox necessary for maintaining balance in our lives. Forward-thinking organizations are beginning to understand the importance of digital well-being and are taking proactive steps to help their employees find a healthier balance between screen time and offline life.

They are also encouraging employees to take regular offline breaks. Automattic, the company discussed at the start of the chapter, offers its employees flexibility to create schedules that allow for long walks, meditation sessions, or whatever form of nonscreen activity helps them recharge. The idea is to recognize that productivity isn't about being online and available 24/7 but about producing quality work and maintaining a healthy state of mind.

To further support their workforce, some companies offer workshops and training sessions focused on managing time in a digital world. The workshops cover topics such as setting boundaries around screen time, using technology more intentionally, and offering techniques for minimizing digital distractions. Google's "Digital Wellbeing" initiative, for example, provides resources and tools to help users find a better balance with technology, emphasizing the importance of taking time to disconnect.

Of course, these initiatives make sense when your workforce is all in one place where you can easily monitor their behavior and productivity, but what about fully remote workers who aren't physically present? You could introduce "no expectation" periods when employees aren't required to respond to work communications, or team hours where workers across time zones can all be in the same space for a period of time to boost collaboration. To inspire employees to pursue hobbies away from screens, encourage physical activities through company-sponsored fitness challenges or send care packages with books or do-it-yourself kits. Atlassian introduced "ShipIt Days" to its remote workforce to work on any project they choose for twenty-four hours as a creative break from regular work tasks. Or you could simply encourage employees to put limits on their own screen time, especially before bed, by sharing stats on how it affects their health and providing an alternative plan.

Research supports the notion that taking time away from screens can lead to increased productivity and well-being. A study published in the *Journal of Applied Psychology* found that employees who took short breaks throughout the day reported higher levels of job satisfaction, reduced emotional exhaustion, and a greater ability to concentrate.[23] The breaks, particularly when spent away from screens, were associated with enhanced creativity and problem-solving abilities.

The Future of Digital Well-Being

Companies are dealing with well-being challenges around screen time, walk breaks, and hydration, but what happens when new issues arise based on emerging technologies that they haven't yet fully navigated?

For instance, what is the protocol for an employee using VR who reports dizziness, nausea, or eyestrain? What about the stress of securely managing digital identities and accommodating clients across various time zones? What about remote workers' feeling of isolation?

These evolving mental health challenges can induce anxiety and stress, particularly among late adopters wary of being left behind, as well as early adopters grappling with the psychological effects of prolonged virtual immersion. Such diverse experiences underscore the necessity for a personalized approach to employee well-being, tailored to individual levels of technology engagement and job requirements.

As cognitive overload becomes a tangible concern for employees navigating intricate new interfaces and juggling interactions across numerous digital platforms, companies can offer comprehensive training and allocate sufficient time for employees to acclimate to these novel systems, ensuring a smoother transition. Ensuring accessibility for all employees, including those with disabilities, transcends legal obligations, embodying a moral commitment. The development of the technologies must prioritize inclusivity from the outset.

Proactively addressing well-being challenges facilitates a more seamless integration into new digital work environments, protecting a company's most valuable asset—its people. These emerging technologies demand our vigilant care and consideration for the comprehensive well-being of employees. It is your job as a leader to make sure these advancements in technology enhance rather than compromise the quality of work lives.

A New Model for Employee Experience Design

Since the inception of formal employment structures, employee sentiment has been a critical factor in driving improvements in both workplace productivity and organizational culture. In the era of the Industrial Revolution, workers were largely voiceless. The name of the game was output, and human considerations were an afterthought. A groundswell of collective action led to the formation of labor unions, which fought vehemently for more humane conditions—outlawing child labor, establishing fair wages, and institutionalizing employee rights.

In the twentieth century, breakthrough research, like the Hawthorne effect, shed light on the vital connection between a conducive work environment and employee performance. The physical workspace began to evolve: office lighting improved, and people began to have their own desks and workspaces. Yet, with all this newfound concern for the individual, the concept of "discretionary effort" emerged. This was a period of extrinsic motivators: loyalty was often rewarded with a gold watch after decades of service, and failure was met with punitive measures—a proverbial carrot-and-stick approach.

Then came the advent of employee engagement. Tech giants like Google and Apple redefined the game by offering far more than just monetary compensation. Game rooms, endless snacks, hot yoga sessions, and on-site childcare became the new norm. Yet, in this quest for better engagement, many organizations missed the mark. Did anyone actually pause to ask what employees genuinely wanted or needed?

Now we are entering the age of employee experience design, a philosophy that modern HR departments embrace. By using techniques like journey mapping and persona development, HR aims to understand the myriad of needs of a diverse workforce, from inception to exit and beyond. But here's where corporate structures plateau: even the most well-executed employee experience initiatives have limitations for two reasons. First, the existing structure of traditional employment intrinsically carries pain points that cannot be fully eradicated. Second, emerging technologies like web3, AI, and decentralized platforms are accelerating at a pace that leaves even the most agile corporations struggling to attract top talent.

The resurgence of unions is a compelling indicator that something is fundamentally amiss in the modern labor landscape. While unionization once served as a crucial instrument for worker rights, its comeback can be viewed as a sign that current workforce practices aren't adequately addressing the diverse and complex needs of employees. If we aim to attract and retain a high-caliber workforce, nostalgia for old labor practices may not be the answer. This leads us to our critical proposition—beyond the carefully crafted and thoroughly researched strategies of employee experience design, we believe the future lies in decentralization.

Decentralization inherently aligns with the qualities the workforce values most: autonomy, flexibility, and a direct stake in one's work. It offers unparalleled adaptability, allowing for rapid crowdsourced solutions to problems and organic bottom-up innovation. In essence, it provides a dynamic, equitable, and truly collaborative environment that doesn't just adapt to change but thrives on it. This engenders a sense of engagement and responsibility unmatched by any Ping-Pong table or free lunch.

Our Digital Humanity at Work

Digital tools make our work lives more flexible and personalized. We're no longer tied to a physical office. Whether we're working remotely from a mountainside cabin or after a parent-teacher conference, the digital age offers us choices that make work more aligned with our individual lives.

It's easy to assume that the growing role of digital technologies in the workplace could make our jobs feel more mechanical and less human; however, digital transformation is reshaping our relationship with work in ways that make it more meaningful and inherently human.

The power of digital tools allows for a reconfiguration of work that puts people at the center. We're not just talking about convenience or efficiency here—though those are undeniable benefits—but about a deeper, more fundamental change that enables us to connect meaningfully with our work and each other. Companies are increasingly embracing this shift, looking beyond simple productivity metrics to consider employee well-being as a critical performance indicator. According to a Gallup poll, companies with high employee engagement are 21 percent more profitable than those without.[24]

As digital workspaces continue to evolve, who knows what other opportunities for meaningful work might emerge? We might soon find ourselves in a metaverse office, collaborating in real time with colleagues and AI agents from around the globe, using avatars that allow for new, creative forms of self-expression. But even as we venture into these exciting new territories, the end goal remains the same: to create work environments that are both efficient and genuinely fulfilling.

The future of work is digital and wonderfully human. It's a future that understands the importance of both individual well-being and collective progress, driven by technologies that enhance our human capabilities. And it's a future that we are already starting to live today.

As we continue this journey into the digital age, let's not lose sight of what's truly important: building workspaces—whether physical or virtual—that allow us to connect more fully in ways that bring about the richness of what makes us human. Let's keep our focus on what truly matters: creating work environments that cultivate deeper connections and embrace the full spectrum of our humanity. These spaces should not only facilitate our work but also enrich our lives, reminding us that true progress is achieved by unlocking our collective potential.

CONVERSATION STARTERS

1. In light of the significant preference for hybrid work setups highlighted by the 2023 Gallup report, what strategies can your company implement to ensure flexibility while maintaining productivity and connectivity among team members?
2. What measures can your organization take to ensure fair advancement opportunities for all employees regardless of their physical work location?
3. In experimenting with metaverse platforms for work-related activities, how is your organization balancing the innovation with considerations for employees' well-being and preventing overreliance on virtual interactions?
4. How is your company adapting its policies and infrastructure to support employees who choose a digital nomad lifestyle, ensuring they remain integrated and productive?

CHAPTER 11

Credentialing, Upskilling, and Continuous Learning

In the fall of 2022, amid the frenzy of college applications, Deborah's son Drake found himself questioning what those around him—and society—had often told him was the path to career success. With a strong academic and athletic record, the path to an Ivy League school or the US Naval Academy seemed clear. Yet Drake questioned the true value of this conventional path, feeling disconnected from the promise of happiness and success it seemed to offer. Soon he chose an unorthodox route and decided to postpone college indefinitely to explore a life defined by his passions and values, not by societal expectations.

Drake's journey of self-discovery took him away from the structured life he knew, eventually landing him in Brazil. There he leveraged his skills as a budding documentary filmmaker, trading cinematic storytelling for room and board at the SOS Mata Atlântica Foundation, a nonprofit dedicated to reforestation. This experience was just the beginning of Drake's quest to live authentically, free from the shackles of the algorithm that had defined so much of his online existence.

As Drake navigated the uncertainties of life abroad, he encountered a community of like-minded digital nomads. These encounters opened his eyes to the potential of leveraging digital platforms for a living, far removed from the traditional corporate life. Inspired by these new friends, Drake immersed himself in the digital-nomad lifestyle and embraced the freedom and adaptability it offered.

During these formative experiences, Drake stumbled on a pathway to build the skills and credentials necessary for success on his own terms. The world of blockchain, cryptocurrency, and DAOs promised a new educational promise where learning was relevant and directly applicable to the emerging digital economy as it aligned with his passions.

Drake's decision to forgo traditional higher education in favor of developing real-world experience and expertise represented a bold bet on his future career and happiness. He became a proponent of decentralized learning, earning credentials and certificates that were recognized across the globe and emblematic of the cutting-edge skills in demand. Drake's journey illustrates a broader shift in how we perceive education, career development, and success in a rapidly changing digital landscape. This story is a testament to the transformation of work in realigning the expectations of today's youth with the skills they hope to develop and the passions they wish to pursue.

Is the College Degree Still Relevant?

A student in the late 2010s might have thought that learning how to code was the golden ticket to a secure future. The student gets into the best university she can afford and then spends nights and weekends grinding through classes in Python and JavaScript frameworks, all along fantasizing about her future as a software developer at a big tech company. Today, that path is much less certain. Gen AI systems can write code, debug, and even optimize algorithms better than many human developers. Where does that leave students and their coding skills?

Take a moment to observe the inefficiencies and shortcomings of the university education system, which have convinced us that four years and an average of $104,000 is the only way to land a decent job.[1] That's just not true anymore. In a 2017 study led by researchers at Harvard Business School, the report "Dismissed by Degrees" found that more than 60 percent of employers rejected otherwise qualified candidates in terms of skills or experience simply because they did not have a college diploma.[2] One of the researchers' most revealing findings was that millions of job postings listed college degree requirements for positions that were currently held by workers without them. For example, in 2015, 67 percent of production supervisor job postings asked for a four-year college degree, even though just 16 percent of employed production supervisors had graduated from college. Many of these so-called middle-skill jobs, like sales representatives, inspectors, truck drivers, administrative assistants, and plumbers, were facing unprecedented degree inflation. "That report was a wakeup call for companies, but it definitely took some time to get out there," said Elyse Rosenblum, the founder of Grads of Life, a nonprofit that backed the study and encourages businesses to adopt more diverse hiring practices.

Today, private employers such as Apple, Google, IBM, and Accenture are dropping the four-year college degree requirement for some jobs.[3] Smaller companies, too, are reevaluating the old standards.[4] States like Massachusetts and Minnesota are passing legislation to drop the degree requirement on certain job listings. Companies are increasingly recognizing that skills and competencies offer a more reliable indicator of an individual's potential in a job than advanced degrees. Skills over experience—this recognition is bolstered by research, such as the study coauthored by Nobel Laureate in Economics James Heckman, which emphasizes that personality traits like perseverance, diligence, and self-discipline are crucial determinants of success, often more so than academic grades or even IQ levels.[5] In this evolving environment, employers are favoring skills-based roles and are more inclined to hire passionate individuals who pursue alternative educational routes, such as free online courses and on-the-job experience, instead of conventional higher education.

The shift is partly driven by the urgent need to fill positions from a diminishing talent pool, which has led to an increased appreciation for practically trained employees. These workers often bring valuable experience and practical knowledge that can surpass the theoretical teachings of traditional academia.

This trend challenges the historical overemphasis on formal qualifications and highlights the growing importance of practical skills and continuous learning in the modern workforce, making them fundamental to professional development and the success of organizations. Why, then, are many companies still relying on old standards when it comes to finding new talent? How can leaders ensure that they are adopting these new technologies to remain relevant?

The Great Skills Reset

A study conducted by experts from SHRM and the Burning Glass Institute highlighted a shift they termed "an emerging degree reset" in the hiring landscape.[6] After examining over fifty-one million job listings from 2014 on, the study revealed that between 2017 and 2019, approximately 46 percent of jobs categorized as "middle-skill" and 37 percent of those labeled "high-skill" no longer specified a bachelor's degree as a requirement.[7] The postings focused on applicants' technical and social skills instead. The researchers estimated that, if current trends continue, about 1.4 million additional jobs could become accessible to individuals without college degrees within the next five years. "Jobs do not require four-year college degrees," the report's authors wrote. "Employers do."

But Rosenblum argues that change isn't always so easy. One of the biggest barriers is just changing mindsets: "Getting more employers to rethink their degree requirements will take hard work. Employers have grown up in a system where the four-year degree is the proxy and there's a perception that it's risky to do something different."[8]

As we have argued throughout this book, the advancements in gen AI are upending traditional career paths and learning models, forcing a redefinition of preparing for the future of work. We're not talking only about job displacement; we're talking about the evolution of how we learn, adapt, and stay relevant in a dynamic ecosystem that's increasingly shaped by emerging technologies. DAOs, smart contracts, and blockchain technology are making the workplace and learning more open, secure, and fair, helping people keep up with new skills and stay relevant in a fast-changing job world.

In a future that's constantly being rewritten by algorithms; leaders must shift their mindset. Organizations must adapt to the new ways the workforce is gaining knowledge and experience, and ensure continuous learning once individuals are hired. This kind of learning does not involve cramming facts and figures the night before an exam. It involves a more fluid, ongoing process of skill acquisition using the opportunities offered by advanced technologies.

New Forms of Education

With new technologies come exciting opportunities for individuals to develop skills and meet the demands of a competitive workforce. While we all may be familiar with online courses and remote classes, new platforms are taking advantage of web3 and AI to offer more inclusive access to learning and skill building. As you look to understand the skills that applicants are bringing to the table, here are a few pertinent developments to keep in mind.

Credentialing beyond Degrees

When skills outweigh traditional degrees, blockchain offers a new way to showcase expertise and accomplishments. These digital badges or credentials are not only secure but also easily verifiable, reducing the risk of false information to virtually zero. According to a PwC report, 75 percent of HR professionals believe that blockchain will be a standard method for credential verification by 2029.[9]

Imagine a software developer with a diverse skill set. They've not only completed a variety of courses but have also actively contributed to open source projects and online developer forums. In the past, their résumé might have been limited to listing a formal degree and perhaps a few notable experiences described in brief bullet points—not much to go on if you're trying to understand the breadth of this person's skills, capabilities, and experiences.

But with blockchain-based credentialing, capabilities are no longer confined to bullet points. They are displayed in a more nuanced, comprehensive manner, as shown in figure 11-1. An individual can compile

FIGURE 11-1

How decentralized identity works

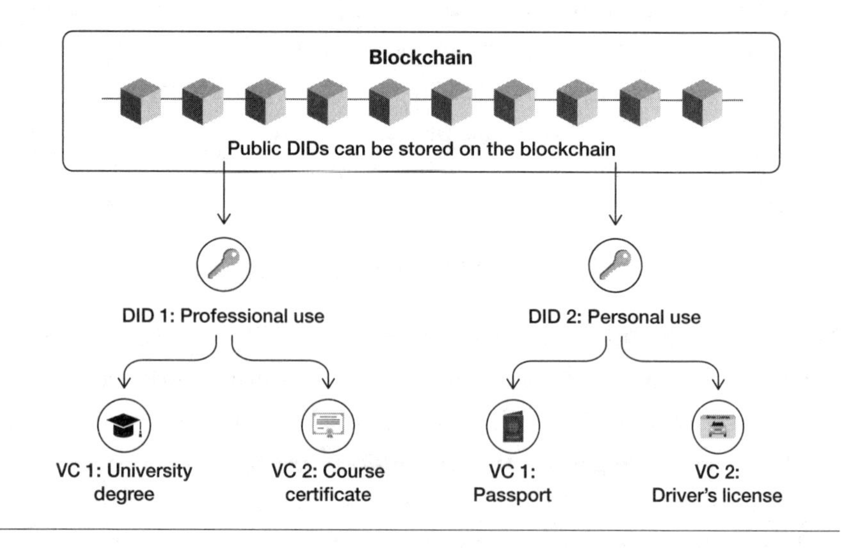

rich, dynamic credentials that reflect their real-world skills and accomplishments. They can even include proof-of-work links that showcase the actual projects they've completed, verified on the blockchain for added authenticity. For employers and clients, this provides a far richer, more reliable measure of someone's true abilities, effectively changing how expertise is proven and recognized. The future of credentialing is decentralized, secure, and tailored to showcase your unique skills—aligning with the career demands of the twenty-first century.

A Decentralized Playground for Learning

Let's imagine a model of learning where participants are rewarded for acquiring new skills and knowledge in the moment when they acquire them. With the advent of decentralized technologies, learners can now directly earn from their educational pursuits without the delay of financial rewards.

This learn-to-earn model offers a direct incentive for learning, making education as much a pathway to potential future earnings as it is a source of immediate value. The model encourages continuous learning and skill development by financially rewarding learners as they progress. It's particularly transformative for lifelong learners and those in underserved communities, as it provides access to education and training that might have been financially out of reach. Ultimately, the learn-to-earn model democratizes education by aligning the incentives of learning with the tangible benefits of earning, thereby motivating individuals to stay relevant and adaptable in their professional lives.

For instance, BitDegree is a platform that is pioneering the use of token-based incentives and blockchain technology in the educational sector by employing a decentralized approach to recognize and reward educational achievements. When a student completes a course or a milestone, they receive tokens, which they can exchange for other courses or even real-world goods and services. Imagine the power of this self-sustaining ecosystem to motivate learners and continue their educational journey.

What sets this approach apart is the use of blockchain technology to record achievements. Unlike traditional transcripts or certificates, which can be lost, damaged, or even falsified, blockchain provides a tamper-proof permanent record. This transparent verification system certifies that the accomplishments are secure and easily shareable, which can be a tremendous advantage in both academic and professional settings.

The impact is particularly profound when we look at it through the lens of social equity. Traditional educational systems often require significant financial investment, which creates tangible barriers for individuals from less privileged backgrounds. Platforms like BitDegree and Degreed could dismantle these economic barriers by paving the way for a more inclusive and accessible education and, by extension, career advancement.

This model also opens the door for global collaboration and knowledge sharing in ways that were previously unimaginable. Someone in a remote village can learn software development from a top expert in Silicon Valley, while a young woman in an urban center can acquire sustainable farming techniques from a seasoned professional in rural Africa.

Tailored Education for the Digital Age

Gen AI can act like a personalized tutor by curating educational content that matches an individual's unique learning style and offers targeted exercises, even generating practice questions in real time based on their progress. Imagine a virtual career coach that can assess your skills, track your career aspirations, and guide you through a personalized learning path.

According to a 2024 report from eLearning Industry, AI in the education market is expected to grow to $10.38 billion by 2028, demonstrating the increasing investment in personalized, AI-driven learning experiences.[10] For instance, language learning apps like Duolingo utilize AI algorithms to create custom lessons that adapt to one's skill level and learning pace. Similarly, AI-powered platforms like Coursera and LinkedIn Learning not only recommend courses based on a learner's professional history but can also identify gaps in their skill set that they might not have noticed.

This dynamic, adaptive approach to learning has proven effective in sustaining learner engagement and motivation. Research from the MIT Media Lab indicates that personalized learning increases learner engagement by up to 50 percent.[11] The goal is to enrich the educational experience, not just to hold interest. For example, AI-powered career coaches can help identify microcredentials or specific skills that are in demand in an individual's chosen field, thus steering them toward more lucrative and fulfilling career opportunities. These platforms interact with a user in the same way a human tutor gives real-time feedback and guidance. In a world increasingly influenced by gen AI, the ability to adapt and learn new skills will become more than a luxury—it will be a necessity for career growth.

Lifelong Learning: The New Corporate Currency

So far, we've looked primarily at bringing new workers *to* your company, but in this era of constant disruption, the notion of "completing one's education" is antiquated. According to a Gartner report, companies that support a "culture of continuous learning" will have employees who are 37 percent more productive.[12]

Consider Linda, a marketing executive in her mid-fifties, who once felt secure in her career trajectory. With the advent of data analytics, narrow AI algorithms, and automated marketing platforms, she realized that her skills were at risk of becoming outdated. Faced with a choice between resisting technological advancements and becoming obsolete, or adapting through continuous learning, Linda wisely chose the latter. She enrolled in online courses covering topics such as data analytics and specialized AI in marketing. Now, Linda stands as a key player in her organization by successfully merging time-honored marketing practices with cutting-edge tools.

The skills that were once sufficient for a thriving career are continuously evolving, thanks to rapid technological advancements. This brings us to a critical point: the emergence of new types of learning that are indispensable for staying relevant in today's workforce.

Adaptability as the New Currency

Adaptability has become the new currency in the job market—a shift primarily driven by the rapid pace of technological advancement. The concept of acquiring a set of skills once and leveraging them throughout a career is becoming obsolete. Instead, adaptability—the ability to learn, unlearn, and relearn—is taking center stage, demanding a radical rethinking of how we approach professional development and growth.

According to a report by the World Economic Forum, 54 percent of all employees will require significant reskilling by 2029.[13] This dynamic environment requires that individuals embrace continuous learning and upskilling to cultivate an agile mindset capable of navigating the complexities of modern industries. The onus is on organizations to create environments that support and promote a culture of adaptability. Such cultures are characterized by ongoing educational opportunities, regular skill assessments, and pathways for employees to explore new technologies and methodologies. They encourage a mindset that welcomes change and learning as opportunities for growth and innovation.

The ability to rapidly adapt and acquire new skills becomes a critical factor in career progression and organizational resilience. Companies that invest in a culture of adaptability, backed by innovative systems like those offered by AI and web3, position themselves to thrive amid technological evolution. The essence of adaptability as the new currency is a symbiosis between continuous learning, upskilling, and the transformative potential of web3 credentialing.

As you work to upskill your workforce and look to disruptive technologies and web3 models, consider the following new ways of learning:

Learning in Bytes: The Rise of Microlearning and Just-in-Time Knowledge

The traditional education and training approaches, with extended coursework and long-duration degrees, often fail to align with the fast-paced demands of the modern work environment. Microlearning has emerged as a compelling alternative, offering bite-size, easily digestible lessons that you can integrate seamlessly into your daily work routine. Imagine a project manager who suddenly needs data analysis skills; instead of committing to a yearlong course or off-site training program (potentially leaving work for an extended period), they could engage with a series of focused microlearning modules. The modules allow the project manager to acquire essential knowledge and immediately apply it, which enables them to learn without disrupting their career trajectory.

Learning Communities: The New Collaborative Classrooms

Various online learning communities facilitate free knowledge flow among learners from diverse backgrounds and geographical locations. Take the example of coding boot camps. These short-term, intensive programs offer technical training and provide a sense of community. According to a Career Karma report, the coding boot camp market grew by 4.38 percent year over year, with 33,959 students graduating from online coding boot camps alone.[14] Forums within these communities provide peer-to-peer help, mentor guidance, and knowledge sharing with alumni, making it a complete learning ecosystem.

Gamified Learning: Engaging through Play

Gamified learning has been a powerful tool for decades to captivate and educate simultaneously. By integrating the mechanics of games into the learning process, this approach transforms education into an engaging, interactive experience. Learners might embark on quests to solve complex puzzles related to their field of study or compete in leader boards by completing educational challenges. For example, language learners could improve their skills by navigating through a game that simulates real-life conversations and scenarios, earning points and rewards as they progress. Gamification taps into the intrinsic motivation of learners, encouraging them to explore and learn without the pressure of traditional assessments. This method significantly increases retention rates by presenting information in a dynamic and interactive context that is more enjoyable. As digital platforms evolve, gamified learning will continue to redefine educational engagement.

Experiential Learning: Real-World Application and Simulation

With the rise of immersive technologies like AR/VR, experiential learning stands out as a transformative educational model by focusing on learning by doing and emphasizing the application of knowledge through real-world simulations and interactive scenarios. Picture a medical student practicing surgical techniques in a risk-free VR environment or a marketing professional experimenting with different campaign strategies in simulated market conditions. These platforms enhance the depth of understanding and significantly improve retention rates by engaging learners in active problem-solving and decision-making processes. With the advancement of VR/AR technologies, experiential learning is set to redefine professional training and education by bridging the gap between theoretical knowledge and practical application, making it an indispensable asset for the future workforce.

The future of learning is both flexible and communal, facilitated by the innovative use of technology. The emergence of microlearning modules and online learning communities means that acquiring new skills or pivoting in your career doesn't have to be a monumental, time-consuming endeavor. You can learn at your own pace and on your terms, but also with the support of a global community. This new reality of lifelong learning is tailor-made for an era marked by rapid technological advancements and constant change.

Embracing Uncertainty: The New Social Contract

As a leader in your organization, developing a culture of continuous learning, unlearning, and relearning is key to building career adaptability. The very technologies that challenge traditional career frameworks present opportunities for growth and resilience. Embracing these changes with openness and encouraging your workforce to do the same can transform potential disruptions into powerful avenues for professional development and organizational excellence.

Therefore, don't get too wrapped up in entering an era of doom and gloom where robots and algorithms take all our jobs. Rather, take courage knowing that we are crossing the threshold into a future rich with opportunities for those willing to stay agile and open to learning. In this intricate landscape of rapidly evolving technology, new skill sets, and continuous learning, embracing uncertainty is a blueprint for thriving in a world of endless possibilities.

The future of work is taking shape at the intersection of rapid technological advancements and evolving societal expectations. Traditional career paths are undergoing radical transformations, and a nine-to-five job will become more an exception than a rule. This ever-changing landscape calls for a new social contract—a flexible, dynamic agreement between workers, employers, and governments that empowers individuals to thrive in their careers.

Lifelong Learning Accounts: Democratizing Access to Education

Picture a corporate environment where every individual has a dedicated fund specifically for ongoing education and skill development. Contributions could come from multiple sources—individuals themselves, their employers, and potentially even governmental bodies. Lifelong Learning Accounts (LiLAs) are employee-owned educational savings accounts that help pay for education and training expenses, which could be a financial and motivational catalyst to encourage people to invest in their own development. This could have an impact particularly on those in transitional phases of their careers or workers in industries that are rapidly evolving. By democratizing access to continuous learning, the accounts aim to make it easier for all people to adapt, grow, and thrive in the ever-changing world of work.

France's *compte personnel de formation* (CPF) is a great model in the realm of continuous learning.[15] This personal training account empowers workers by letting them accumulate training credits on an hourly basis. They can use the credits for a diverse array of courses, from vocational training to advanced professional development. Whether you're a seasoned professional looking to update your skills or just entering the workforce, the CPF system caters to a broad spectrum of educational needs. The initiative is not just limited to traditional employment sectors; it's adaptable to the evolving landscape of gig work, freelancing, and other nontraditional job roles. The success of CPF underscores the potential for such accounts to be scalable and widely adopted. It is a compelling example of how a well-designed, flexible system can democratize access to education and training, thereby fostering a culture of lifelong learning.

Ownership of Data and Identity: The Web3 Revolution

The concept of a "portfolio career" is gaining traction—individuals diversify their skill sets across multiple domains. This shift indicates a broader trend: away from an age of specialization to an era when having a varied skill set could be a person's most valuable asset. In this new landscape, flexibility and adaptability are essential qualities.

Platforms like uPort or Sovrin, for example, utilize blockchain technology to provide self-sovereign identities, allowing users to manage and control access to their personal information. This step toward empowering individuals in the digital realm gives them the autonomy to decide who can access their data and for what purpose.

Web3 technologies are fundamentally shifting the narrative around personal data ownership. In contrast to the current web landscape, where data monetization largely benefits big tech companies, web3 offers individuals the tools to take control and become active stewards of their own data.

Ethical AI and Human-AI Collaboration: Partners in Progress

As AI increasingly weaves itself into the fabric of our work lives, the importance of aligning this technology with human ethics and values is paramount. Is the rise of AI in the workforce a threat to human jobs? Maybe as we know them today, but emerging generations have an unprecedented opportunity for symbiotic collaboration between humans and machines.

The collaboration is about augmenting human capabilities, enhancing decision-making, and even fostering creativity. AI can handle data analysis, freeing up humans to focus on strategic planning or creative problem-solving. It can also assist in real-time decision-making, providing insights that might not be immediately obvious.

The Soft-Skills Paradox: AI and Emotional Intelligence

As technical skills are increasingly automated, soft skills have come to the fore as highly valuable assets. According to a Deloitte study, soft skills, which include traits like communication, critical thinking, and cultural awareness, will be required for two-thirds of all jobs by 2030.[16]

While gen AI is a powerhouse for enhancing hard skills, can it truly teach us the subtleties of emotional intelligence (EI) or the complexities of human behavior? As we automate various aspects of work and learning, the distinctly human qualities—such as teamwork, empathy, and leadership—become increasingly vital.

Maintaining soft skills and EI is important in humanizing our interaction with new technologies, which cannot yet replicate EI. Skills, such as empathy, communication, and creativity, alongside technical abilities, are deemed essential for maintaining meaningful human connections at work in an increasingly automated world.

Even as we automate many elements of work and learning, the human element—teamwork, empathy, and leadership—becomes more important than ever. Schools and online platforms have started to recognize this, incorporating soft-skills training alongside technical courses. But the most effective training might still be human to human: mentorship, real-world teamwork, and leadership experiences that no AI can replicate.

Facilitating Learning as a Strategic Investment

For too long, modern society has been organized around a three-stage linear model of education, which starts with grade school and higher education, followed by work and hands-on experience, and ends with retirement. This model no longer serves workers. Pessimists highlight that the technological changes will eliminate jobs and push other workers to a new, unfamiliar place of lower wages and stagnant opportunities. The optimists, however, see disruptive technologies as a unique chance to reshape the educational system, making learning more personalized, accessible, and lifelong.

As emerging technologies like blockchain technology, AI, smart contracts, and decentralized systems go mainstream, credentialing is on its way to becoming more personalized, transparent, and accessible to everyone. The tools like lifelong learning records, microcredentials, and peer-to-peer validation tailor learning to individual needs and global acknowledgment. By leveraging web3's infrastructure, we're moving toward a future where high-quality education is universally accessible and individuals can reliably showcase their skills in a way that's trusted and easily verified.

Your role in facilitating learning takes shape in a more proactive model that is infused with care and consideration for each individual's learning journey. By creating a culture that prioritizes learning and development, you are investing in the skills of your current employees and shaping the work environment and expectations for future generations of workers. Investing in learning is good for business—and essential for long-term success and survival.

CONVERSATION STARTERS

1. How do you change your mindset about hiring requirements and reconcile the prestige and structured learning of traditional education with the flexibility and real-world applicability of self-directed learning paths?
2. Where in your company is upskilling needed most urgently? What strategies can you use to ensure that employees are also motivated to engage in lifelong learning?
3. How can web3 technologies facilitate personalized learning experiences that align with individual career goals and industry demands?
4. What challenges and opportunities do decentralized credentialing systems present for job seekers to demonstrate their skills and competencies? How can the systems change the way we assess potential hires?

CHAPTER 12

People Operations in the New World of Work

Tracie Sponenberg was a unique kid; she always knew from an early age that she wanted to be in human resources. Her aunt worked in a field called "personnel," and she was fascinated. The creation of policies. Enforcing rules. Hiring people. Firing people. After earning a degree in industrial-organizational psychology and entering the workforce, she quickly became quite proficient at her job, which haunted her years later. Sponenberg recalled, "I remember distinctly the first time I fired someone. I was easily able to disconnect and end that person's career without emotion. Many terminations later, my boss told me, 'You are really good at that. And I'm not sure that's a good thing.'" It earned her the reputation that you often hear about—if HR is coming, someone must be getting in trouble or fired.

As the personnel function transformed over the years into human resources, Sponenberg began to rethink her long-held beliefs. A particularly impactful exchange with a team member is an example: "Someone on my team had an incredibly difficult conversation with me and told me that she had a problem. . . . and it was me. I was not giving her the time and attention needed, nor was I giving her feedback. She changed my life that day." Sponenberg went on to become the senior vice president of HR at the Granite Group, where she found her rhythm working with the CEO to lead with a people-first approach. The result was a completely overhauled HR department that became "people operations," and Sponenberg became its first chief people officer.

In an interview with the authors she said, "People are not resources. They are complex individuals with emotion and passion. We place an incredible focus on the individual, guiding them through their unique experience from before they are a candidate through the time they leave the company. If there is a CFO at a company, there needs to be a chief people officer. So we made the change. We want our people to understand that we are there for them."

Sponenberg's transformation is more than just semantics. It reflects HR's expanded purview, which now stretches well beyond processing payroll, benefits, and pay increases to encompass areas traditionally reserved for the upper echelons of management, including talent development, leadership training, change management, and culture initiatives. Many other companies are embracing the philosophy that workers are not resources to manage but individuals to empower.[1] It's become clear to CEOs everywhere that HR's role is evolving beyond its base function and has become more crucial now than before. And as disruptive technologies take a larger role within organizations and in the lives of employees, we pay special attention to the human side—the people we're working with.

A New Role for HR

We're standing at the crossroads of traditional HR practices and the untapped potential of a decentralized, tech-savvy ecosystem. HR must now lead in discovering how we leverage these advancements to enhance People Operations and champion a more dynamic, inclusive, and flexible work environment.

A lot of what's in an HR practitioner's toolbox is outdated—annual performance reviews, set salary scales, the same old training programs, and so on. Those are the familiar comfort food of HR. It's easy to reach for the familiar, but that is not necessarily what's needed in this new context. It's time to be brave, think differently, and throw out that old playbook.

HR departments are well positioned to pivot from traditional, time-consuming administrative tasks to focus on strategic initiatives that directly impact the core of an organization's success. In this new landscape, HR will be the designer of people support systems outside geographical restrictions and across time zones. It will craft strategies that leverage blockchain and smart contracts for transparent and equitable management practices, facilitate continuous learning and upskilling to keep pace with technological advancements, and most importantly, nurture a cohesive organizational culture when the office is no longer one physical place.

In short, HR's function will guide organizations through this transition to guarantee that people remain at the heart of the new work ecosystem, especially in the following areas:

Talent development. With administrative burdens automated, HR is devoting more resources to identifying and nurturing the unique strengths of each employee. This involves personalized career paths, mentorship programs, and continuous feedback loops, ensuring that employees are not only satisfied but are also continuously growing and contributing to their maximum potential.

Culture initiatives. The heart of any organization is its culture. HR professionals are now spearheading initiatives that foster a positive, inclusive, and dynamic work environment. By leveraging data and employee feedback, HR can tailor programs that build a sense of belonging, recognize achievements, and encourage innovation and collaboration.

Training. The importance of ongoing training in a rapidly evolving work landscape is obvious, as HR implements cutting-edge learning platforms and microlearning modules, making training more accessible and engaging. This approach ensures that the workforce remains agile and can adapt to new technologies and methodologies.

Onboarding and recruiting. The first experiences of new employees significantly impact their perception and loyalty to an organization. With more bandwidth, HR can craft onboarding experiences that are comprehensive, welcoming, and informative, setting the stage for long-term engagement. Similarly, recruiting efforts can be more focused and strategic, leveraging social media, professional networks, and AI-driven tools to attract top talent that aligns with the company's values and goals.

In essence, the shift toward utilizing emerging technologies allows HR to transcend its traditional roles, placing it at the forefront of driving organizational growth through strategic talent management. This will simultaneously enhance the employee experience and position the company as a leader in innovation and workplace culture. As HR embraces its new role, it becomes a catalyst for transformation, shaping the future of work where people are the most valued asset.

People Operations Strategy Guide

To navigate the emerging work landscape effectively, people operations need a well-crafted strategy. It's crucial to think holistically about how disruptive technologies will intersect with various facets of your business. With a carefully laid-out plan, you can lead your organization into a future where the employee experience is revolutionary.

Upskilling for Web3 Literacy

According to a Deloitte survey, an overwhelming 90 percent of organizations feel that digital newcomers could disrupt their core business, yet only 44 percent consider themselves adequately prepared for the impending digital revolution.[2] The disparity presents a dire need for immediate action. Implementing upskilling programs is one of the most effective ways to bridge this knowledge gap. Consider organizing boot camps that delve into critical areas such as blockchain technology, decentralized finance, and the ethical considerations surrounding smart contracts. By doing so, you're equipping your employees with the essential skills required to thrive in a digital-first ecosystem.

As the work landscape continues to evolve at an unprecedented rate, upskilled employees become invaluable assets to your company and the broader industry. By gaining fluency in web3 concepts and technologies, they are better positioned to adapt to shifts in market dynamics, ensuring a competitive edge. A well-educated workforce contributes to your organization's reputation as a forward-thinking, employee-centric entity, which is invaluable in attracting and retaining top talent.

Think of upskilling as an investment in your human capital—a measure that offers returns in the form of increased productivity, elevated job satisfaction, and a workforce that's empowered to innovate. Given the complexities of emerging technologies, these skills require structured, strategic learning programs. By prioritizing and investing in such programs, you're preparing your team for the inevitable transformations within your organization by enabling them to become key players in the broader, rapidly evolving digital landscape.

AI-Driven, Decentralized Recruitment

One of the most pressing needs will be the recruitment of new talent proficient in the nuances of emerging technologies. AI can be a powerful ally in this endeavor. Speed is crucial in the fast-paced landscape of web3, where securing the right expertise can provide a decisive competitive edge. AI tools can efficiently sift through large pools of candidates to identify those who meet specific skill and experience criteria, thereby significantly reducing the manpower and time traditionally required for initial screening phases. We must be cautious, however, that our AI data has not inherently included bias in its data set, a major problem large language models have been dealing with, as in the example of Google's Gemini depicting historically inaccurate images of people (see chapter 9).

Decentralized identity verification platforms offer another innovative avenue for streamlining recruitment. Platforms like Bloom and Serto provide secure, self-sovereign identity (SSI) solutions, a digital identity system that gives users complete control over their personal information, including who can see it and when, that could revolutionize the vetting process. The potential here is twofold: First, by automating identity and credential verification, decentralized platforms can further accelerate hiring timelines. Second, the platforms offer an added layer of security and authenticity, ensuring that the talent you bring onboard has already been rigorously vetted. When you combine the predictive analytics of AI with the secure verification capabilities of decentralized platforms, you're speeding up the recruitment process and enhancing the quality and fit of your hires. These integrations of AI could very well represent the future of talent acquisition.

Strategic Integration

A compelling study by *MIT Sloan Management Review* and Deloitte underscores that digitally mature organizations are better at integrating their digital strategies within their overarching business goals.[3] This implies that a nuanced approach is vital for a seamless transition into the web3 realm. Rather than an abrupt overhaul, consider a phased approach that complements your existing processes. For instance, you could initiate the transition by moving specific departments or high-priority projects to a DAO. Another tactical step could be piloting a DeFi system for handling expense approvals or payroll, which would provide real-time insights into the efficiencies and potential bottlenecks of a decentralized financial structure.

A great starting point is to organize decentralized teams, or Dteams. Here, teams operate autonomously and leadership is distributed in order to achieve a more democratic approach to decision-making and project management. This model offers a sense of ownership among employees that aligns with the dynamic nature of today's work environments. It allows companies to move with more agility and solve problems more quickly with fewer people.

As you proceed with these incremental changes, continually assess outcomes against predefined metrics and KPIs. This evaluation will help you fine-tune the ongoing integration process by providing valuable learnings that can inform future scaling decisions. The objective is to build a resilient framework that supports both current operations and future innovations. Through a carefully considered, step-by-step approach, you can mitigate risks and better adapt to the challenges and opportunities that come with digital transformation.

Governance and Policy Revisions

The transition to a decentralized organizational model calls for a comprehensive reevaluation of existing company policies and governance structures. This shift can feel like transitioning from driving on the right side of the road to the left (or vice versa). Findings from the World Economic Forum underscore that corporate governance in the age of emerging technologies like web3 remains a largely uncharted territory, ripe for innovative solutions.[4]

HR leaders are now tasked with pioneering policies that accommodate these changes, ensuring that IP and employee rights are protected in environments where the lines of authority are blurred. This demands a reevaluation of traditional governance models to embrace the flexibility required by blockchain and decentralized systems. Questions of accountability, legal compliance, and ethical conduct in such models necessitate innovative approaches to policy development and enforcement.

HR's expertise in organizational behavior, policy development, and compliance, combined with a willingness to collaborate across disciplines, is critical in building governance structures that adapt to safeguard the organization's integrity and its employees' welfare. By steering these efforts, HR leaders play a central role in ensuring that the organization's governance keeps pace with the rapid evolution of decentralized technologies, making HR a cornerstone in the successful integration of the new models into the workplace.

Pilots and Feedback Loops

Some of the transformations may be too large financially or logistically to launch companywide from the outset; therefore pilot programs are invaluable tools that allow your organization to test the efficacy of strategies and initiatives in a controlled environment. The pilots are a litmus test, giving you an opportunity to make data-driven decisions and refine your approaches based on actual experiences and outcomes.

For the pilots to be useful, however, you must establish robust feedback loops for continuous learning and adaptation to help you identify what works well and what doesn't, enabling the fine-tuning of your strategies as you scale. Feedback loops should involve both the executive team and the broader workforce, as they will be the end users of the new technologies. Such inclusivity ensures that the transition to a decentralized model is both effective and aligned with the needs and expectations of your team, thereby enhancing the likelihood of a smooth and successful transformation.

Introducing new technologies into the workplace is about building a culture that is prepared to adapt and thrive in a decentralized environment. Comprehensive planning and foresight are essential, but so are the softer skills of leadership—building trust, facilitating open communication, and nurturing a culture of continuous learning. Your ability to weave together the technological and human elements of your organization will be crucial to the successful transition to a decentralized model.

The rewards are just as much in efficiency gains or cost savings as they are in fundamentally changing the way work happens—creating an ecosystem where contributors are empowered, innovation flourishes, and the work environment itself evolves dynamically. The end result is a stronger, more agile organization that offers a more fulfilling, equitable, and engaging place to work. The horizon of what's possible is expanding, offering a chance to redefine work in a way that aligns with the aspirations of a new generation.

HR Strategies in Action

HR has a great opportunity to shape the future of its organization, and while the people operations strategy guide we propose is comprehensive, we admit that it can also feel overwhelming. If the practical implications of HR's transformation still seem elusive, let's explore the real-world applications of the emerging technologies we've been discussing.

AI-Driven Recruitment and Talent Management

As we noted, AI's capacity for analyzing vast data sets enables HR professionals to identify the best candidates faster and more accurately than traditional methods. AI-powered tools are automating routine tasks, from screening résumés to scheduling interviews, allowing HR to focus on larger strategic initiatives. This evolution signifies a shift toward a more efficient, data-driven approach to talent management.

Companies like LinkedIn and HireVue are at the forefront of integrating AI into the HR landscape. LinkedIn's sophisticated algorithms enhance recruiter tools to make sure job postings reach the most suitable candidates by aligning with their skills and experiences. Meanwhile, HireVue's AI-powered talent assessment suite leverages video interviews to analyze candidates' communication styles to offer deeper insights than traditional résumé screenings. These applications of AI streamline the recruitment process and promise a more objective and comprehensive approach to bringing on new talent, which allows HR departments to focus more on fostering employee growth and driving strategic business outcomes.

Blockchain for Secure Employee Data Management

By creating immutable records of employees' credentials, achievements, and work history, blockchain protects the integrity of data and simplifies the verification process. This streamlines onboarding and compliance processes to empower employees by giving them control over their professional data and credentials.

For instance, Sony Global Education has developed a blockchain-based system for secure sharing of academic credentials, so employers can more easily verify the educational background of prospective hires. Similarly, WorkChain.io utilizes blockchain to automate the verification of work and salary history, reducing the possibility of fraudulent claims and enhancing trust in the recruitment process.

These initiatives underscore the potential of blockchain to revolutionize credential verification and employee data management. Blockchain technology offers unprecedented transparency and efficiency in handling employee information.

Leveraging Smart Contracts for Employee Agreements

As we covered in chapter 7, smart contracts offer HR a tool to manage employment relationships transparently and efficiently, especially when it comes to payroll and paying vendors. Companies like Bitwage are using blockchain for payroll to enable seamless and transparent payment processes across borders without the exorbitant fees and delays associated with traditional banking systems. They have significantly improved the employee experience for remote and international team members.

Chronobank is an innovative platform focusing on the HR sector, particularly in short-term employment contracts in the gig economy. By utilizing smart contracts, Chronobank facilitates transparent and fair work agreements, ensuring timely and accurate payments directly linked to the fulfillment of contract terms.

Likewise, Opolis is a decentralized employment system that uses smart contracts to provide self-sovereign benefits, payroll, and employment status for independent workers. This system allows freelancers and contractors to enjoy benefits typically reserved for full-time employees, such as health insurance and retirement plans, all governed by transparent and enforceable contracts.

The examples showcase how smart contracts are transforming traditional employment agreements into more transparent and equitable arrangements. These companies are leading the way in adopting blockchain technology to enhance the HR function, making the employment relationship management process more seamless for both employers and employees, no matter how they work.

Redefining Engagement and Decision-Making Rights

As organizations adopt more flexible autonomous models of operation, HR professionals will need to devise innovative strategies to keep team members engaged and committed to the organization. As traditional hierarchical structures give way to distributed decision-making and autonomy, the role of HR shifts from overseeing to facilitating. As Tracie Sponenberg, the chief people officer discussed at the beginning of the chapter, describes it:

> HR has evolved from the days of processing payroll, benefits, and pay increases. A lot of the day-to-day work is now (or should now be) done by managers. HR has become People Ops in many places, and we are now coaches, consultants, and guides. There is an incredible focus on the individual, and guiding them through their unique experience—starting before they are a candidate through the time they leave the company—and ensuring that is not only as unique as they are, but as positive as possible.

The move toward a decentralized ecosystem is transforming not just where we work but how work is governed and how employees participate in decision-making processes. This shift has profound implications for organizational dynamics, offering both opportunities and challenges for employee empowerment.

Establishing Collaboration and Community in a Dispersed Workforce

In the virtual expanse of decentralized workspaces, maintaining robust employee engagement and well-being necessitates innovative approaches to foster a sense of community and ensure the holistic well-being of remote employees. The primary challenge in a virtual environment is overcoming physical distance to create a cohesive and engaged workforce. Yet, this distance also opens the door to flexible work arrangements that can significantly enhance job satisfaction and work-life balance. A Gallup poll revealed that 85 percent of employees report higher levels of engagement when given flexible work arrangements, underscoring the potential for decentralized workspaces to positively impact engagement and well-being.[5]

Basecamp has always been ahead of the curve by actively cultivating a culture that deeply values work-life balance. Its commitment to allowing employees the freedom to work in environments where they feel most inspired and productive brings a sense of autonomy and trust, but also highlights its innovative spirit in creating a more engaged and satisfied workforce.

The ever-increasing popularity of AR/VR technologies is revolutionizing the creation of immersive and interactive work environments. By simulating physical office spaces or creating entirely new virtual ones, the technologies can mitigate feelings of isolation among remote employees. Microsoft's Mesh for Teams, for example, integrates mixed reality into daily workflows, allowing more engaging and collaborative virtual meetings.[6] A study by PwC found that VR learners were up to 275 percent more confident to act on what they learned than traditional learners, highlighting the potential of the technologies to not only enhance engagement but also improve training outcomes.[7]

For engagement and retention, personalized career development paths that acknowledge and leverage each individual's unique skills, interests, and contributions are paramount. Engagement initiatives will increasingly rely on technology platforms that support virtual collaboration, recognition, and feedback loops. Despite physical distance, workers will feel connected to their team and valued by their organization.

The Future of HR in a Decentralized Ecosystem

While all the companies we've described show cutting-edge advancements and progress in how HR operates, there's much more to consider if we are willing to use our imaginations. In the not-so-distant future, HR departments will look radically different, embodying the cutting edge of technology while sustaining a deeply human-centric approach to workforce management. Picture an HR department where

AI predicts and fulfills training needs before gaps impact productivity, and where blockchain technology verifies that every contribution and skill enhancement is transparently recorded and rewarded within a truly meritocratic environment. HR facilitators of personal growth and professional development could leverage decentralized systems to empower employees to shape their own career paths.

Envision an HR department where empathy and technology are not at odds but are intertwined, creating a work environment that is both highly efficient and profoundly human. Advanced AI could offer personalized mental health support, recognizing signs of burnout before they become problematic, while blockchain could secure personal data, giving employees control over their own information. Decentralized work models could allow for truly global talent pools, with HR practices that are as inclusive and diverse as the world itself. HR will be able to anticipate changes and craft a workforce that is perpetually ahead of the curve. If HR leaders begin now to integrate these visionary technologies with a deep commitment to human-centric values, the potential for growth is boundless.

Here are just a few ways we envision HR's continual evolution:

HR's seat at the table. HR professionals are now pivotal strategic players, regularly invited into boardrooms to contribute to high-level decision-making. Their insight is critical in leading digital transformation initiatives and driving forward company culture that directly impacts the bottom line. By leveraging AI data and people analytics, HR departments are crucially positioned to navigate the ever-changing work landscape.

AI-powered efficiency. Routine administrative tasks are fully automated, thanks to advanced AI algorithms. Recruitment processes leverage AI to sift through applications, ensuring that candidates are selected based on skill and experience, reducing biases. Sophisticated chatbots that provide instant, accurate responses handle employee inquiries, allowing HR professionals to focus on strategic initiatives.

Hybrid talent management. Decentralization transforms HR into a fluid, dynamic ecosystem. Talent pools are global, with smart contracts facilitating gig- and project-based work alongside traditional roles. This not only expands the talent pool but also introduces flexibility and autonomy in employment contracts, appealing to the modern worker's desire for independence.

Data-driven insights. Data analytics and machine learning offer deep insights into employee satisfaction, productivity, and well-being. HR then uses these insights to tailor benefits, work arrangements, and personal development programs, ensuring each employee's career journey is fulfilling and aligned with their personal goals.

Focus on culture and well-being. HR champions organizational culture and employee well-being in new ways. Wellness programs, mental health support, and community-building activities are not just perks but integral components of the employee experience. VR and AR technologies also bring remote teams together in immersive environments, fostering a sense of belonging and teamwork.

Continuous learning and development. Lifelong learning becomes the cornerstone of the futuristic HR department. With rapid technological advancements, HR curates personalized learning paths for every employee, leveraging platforms that offer microlearning, gamification, and virtual simulations. This ensures the workforce is agile, skilled, and ready for future challenges.

This futuristic HR department is not just a support function but a strategic partner at the heart of the organization. It leads by example, showcasing how technology can be harnessed to enhance human work, creativity, and collaboration.

From Prediction to Practice

You might view our predictions as looming chaos and feel nervous or terrified; we completely understand the fear and see it in the eyes of almost every leader we consult with. But progressive leaders see a playground of potential, a world ripe with opportunity for creativity and innovation for everyone.

When venturing into this uncharted territory, you're not trekking alone. You are part of a vibrant community of forward thinkers, innovators, and future-of-work enthusiasts, all navigating this new landscape together. This camaraderie enriches the journey and emboldens your steps toward embracing transformative technologies. Beyond streamlining operations and enhancing efficiency, you're laying the groundwork for a more engaging, fulfilling work environment.

You have a chance to be a pioneer in redefining the essence of work, making it more fulfilling and human-centric. This is the moment for HR to embrace its brand-new role in crafting a culture where innovation thrives and every individual can achieve their full potential. The path forward may be uncharted, but it's filled with the promise of making work better for everyone.

CONVERSATION STARTERS

1. As you navigate the transition from traditional HR practices to people operations in a decentralized world, how can you ensure that the core of your efforts remains deeply human-centric?
2. In a landscape where technology and automation are increasingly prevalent, what strategies can you employ to maintain, or even enhance, the empathy and understanding that are fundamental to effective people management?

3. How can HR leaders design upskilling and digital literacy programs that are inclusive, ensuring that no employee is left behind in this digital transformation?
4. As the concept of the workplace evolves beyond physical office spaces to encompass decentralized, digital environments, what innovative approaches can HR leaders take to foster a sense of community, belonging, and engagement among remote teams?
5. How can you leverage technology as a means to create meaningful connections and a vibrant company culture that resonates with a global and diverse workforce?

CONCLUSION

The Work3 Transformation Road Map

As we stand on the precipice of a transformative era where AI and other disruptive technologies are set to revolutionize the way we communicate, interact, and build relationships at work, it is imperative that you act now to evolve your organization.

Throughout this book, we have dissected the limitations and shortcomings of traditional employment models—systems that have long prioritized productivity and profit at the expense of human well-being and fulfillment. These outdated structures, characterized by rigid hierarchies, inflexible schedules, and a glaring disregard for the employee experience, have fostered a landscape of widespread disengagement, burnout, and disconnection.

However, the convergence of AI, blockchain, and other disruptive technologies offers us an unprecedented opportunity to rewrite the rules of work. This isn't about incremental change; it's about designing a new paradigm that places human potential at its core. By leveraging these advanced tools to automate routine tasks and streamline processes, we can liberate employees to engage in higher-order thinking, creative problem-solving, and pursuits that align with their innate talents and passions.

The implications of this shift are profound. Imagine a workforce empowered by AI-assisted decision-making, collaborating seamlessly across virtual and physical spaces, their contributions recognized and rewarded through blockchain-enabled smart contracts. Picture a work environment where machine learning algorithms optimize schedules to match individual circadian rhythms, and where augmented reality facilitates immersive, global collaboration.

This human-centric approach to work isn't just about improving employee satisfaction—although that's a crucial outcome. It's about unlocking the full spectrum of human potential, driving innovation, and creating a resilient, adaptable organization capable of thriving in an increasingly complex world.

The time for incremental change has passed. The future of your organization—and indeed, the very nature of work itself—hangs in the balance. Are you ready to pioneer this transformative journey and lead the charge into the Work3 era? The road map lies before you. Let's embark on this revolutionary path together.

Work3 Road Map: A More Human Workplace

Drawing on insights from organizational change management, strategic planning, and innovation, our road map is a blueprint for action and a practical guide that addresses the pivotal questions executives ask: "How can we integrate transformative processes into our operations to minimize disruption while maximizing long-term organizational agility and innovation? How can we harness AI and web3 technologies to enhance productivity and humanize work? What do we need to offer to attract the next generation of talent?"

Having a well-defined road map for a Work3 transition fosters alignment across the organization, abating confusion and resistance to change. Additionally, it facilitates a smooth transition by minimizing disruptions to daily operations and promoting accountability among employees. (Before getting started, see the sidebar "Assemble a Work3 Task Force.")

While traditional employment models may be fading, the spirit of entrepreneurship and continuous upskill development remain vibrant. The core of the Work3 plan is about shaping a new era of work culture that embraces the principles of Work3 and thrives in an ever-evolving landscape (see figure 13-1). The following is the Work3 transformation road map, in twelve steps.

Step 1: Establish a strategic Work3 transformation committee and task force

The first step of your Work3 transformation road map is to establish a lean yet impactful strategic transformation committee comprising seven to nine key stakeholders for a holistic perspective on the transformation's impact across different departments, functions, and employee segments. The committee is a forum for deliberation, decision-making, and problem-solving, enabling stakeholders to address challenges proactively and leverage opportunities effectively. It provides a platform for sharing insights, exchanging best practices, and surfacing innovative ideas to drive continuous improvement and adaptation. This core committee will drive swift, impactful change toward a more agile, inclusive, and AI-enhanced work environment.

The strategic transformation committee should include the following people:

Chief executive officer. The CEO provides the overarching vision for the transformation and ensures top-level buy-in across the organization.

Chief technology officer. The CTO guides the technological integration process, ensuring alignment with the company's overall digital strategy.

Chief human resources officer. The CHRO oversees talent development initiatives and leads the cultural shift necessary for successful transformation.

FIGURE 13-1

A more human workplace: Work3 road map

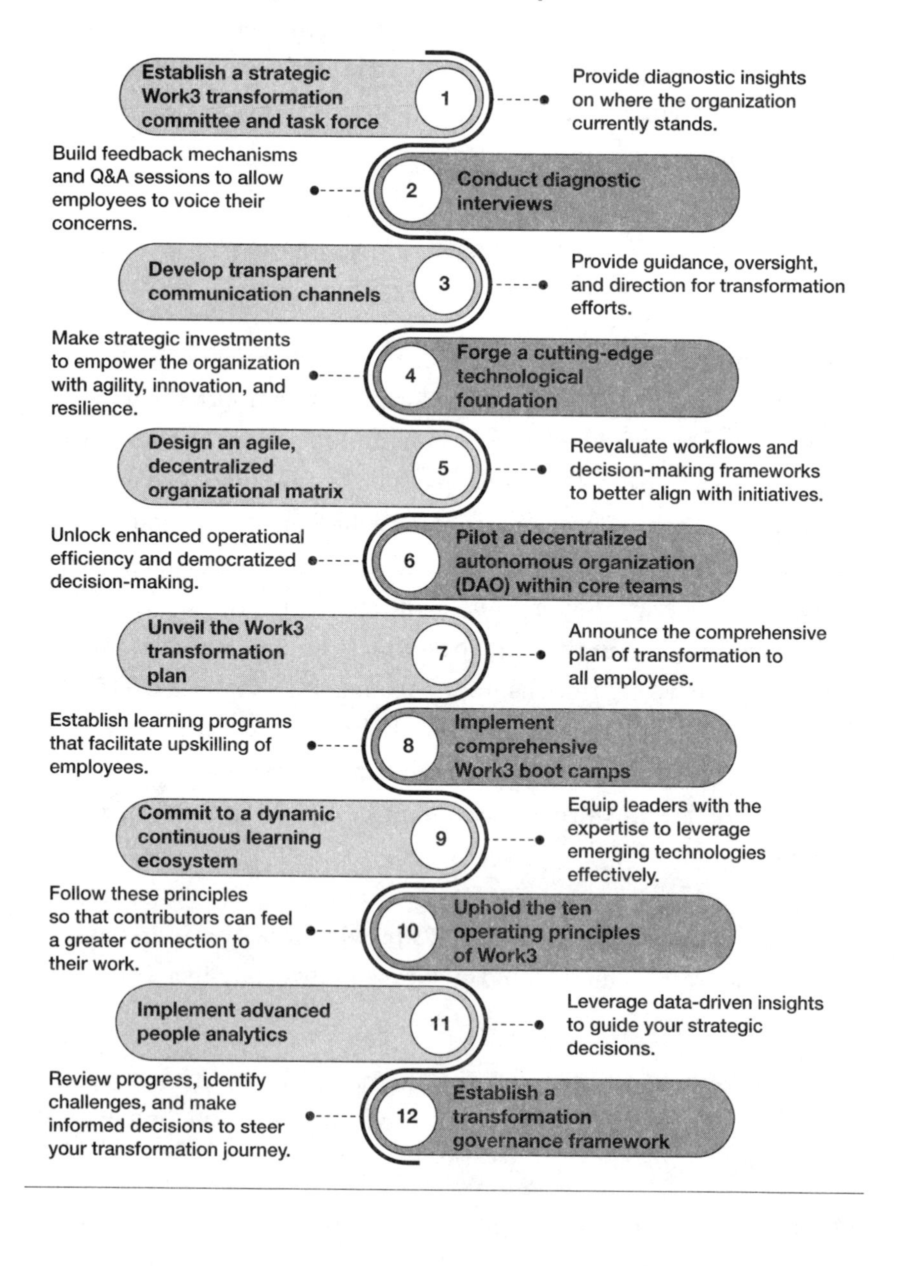
Establish a strategic Work3 transformation committee and task force
1
Provide diagnostic insights on where the organization currently stands.
Build feedback mechanisms and Q&A sessions to allow employees to voice their concerns.
2
Conduct diagnostic interviews
Develop transparent communication channels
3
Provide guidance, oversight, and direction for transformation efforts.
Make strategic investments to empower the organization with agility, innovation, and resilience.
4
Forge a cutting-edge technological foundation
Design an agile, decentralized organizational matrix
5
Reevaluate workflows and decision-making frameworks to better align with initiatives.
Unlock enhanced operational efficiency and democratized decision-making.
6
Pilot a decentralized autonomous organization (DAO) within core teams
Unveil the Work3 transformation plan
7
Announce the comprehensive plan of transformation to all employees.
Establish learning programs that facilitate upskilling of employees.
8
Implement comprehensive Work3 boot camps
Commit to a dynamic continuous learning ecosystem
9
Equip leaders with the expertise to leverage emerging technologies effectively.
Follow these principles so that contributors can feel a greater connection to their work.
10
Uphold the ten operating principles of Work3
Implement advanced people analytics
11
Leverage data-driven insights to guide your strategic decisions.
Review progress, identify challenges, and make informed decisions to steer your transformation journey.
12
Establish a transformation governance framework

Chief financial officer. The CFO ensures the financial viability of the transformation efforts and monitors the return on investment for all initiatives.

Chief operations officer. The COO implements changes across various business processes, ensuring operational efficiency throughout the transformation.

Chief innovation officer. The CIO drives forward-thinking initiatives, fostering a culture of innovation within the organization.

Chief AI officer. The CAIO spearheads the AI strategy and oversees its implementation across all relevant departments.

Employee experience lead. This individual represents the workforce's needs and gathers feedback to ensure the transformation aligns with employee expectations.

External AI/future of work expert. This person offers an industrywide perspective, providing insights on emerging trends and best practices in AI, web3 technologies, and the future of work.

Ultimately, the strategic transformation committee is a driving force behind the Work3 transformation. By combining diverse expertise with a lean structure, this committee ensures a comprehensive, top-down transformation while maintaining the agility needed in today's future-ready, AI-enhanced workplace.

Step 2: Conduct diagnostic interviews

Diagnostic interviews provide crucial insights into your organization's Work3 readiness, laying the groundwork for targeted transformation efforts. As each organization's transformation journey is unique, tailor your interview process to align with your specific goals and context. Take these actions to execute a thorough diagnostic process:

Assemble a Work3 Task Force

The strategic transformation committee provides the overarching vision and leadership for the Work3 transformation. To ensure effective execution, it is complemented by a dedicated Work3 task force—a lean team of six to eight skilled professionals charged with spearheading the day-to-day implementation of the transformation strategy. Working in close alignment with the committee, the task force is the operational backbone of the change effort, translating high-level directives into actionable plans and tangible results.

The task force's roles span program management, technology oversight, people operations, marketing, partnerships, and communications, ensuring comprehensive coverage of the transformation's facets. Members of the task force coordinate training initiatives, equipping employees with the necessary skills and mindset to thrive in a Work3 environment while serving as a dependable resource for guidance and support throughout the transition.

Meeting biweekly, this cross-functional team sets transformation priorities, allocates resources, resolves interdepartmental challenges, evaluates progress, and champions change across the organization. By leveraging AI-powered analytics and agile methodologies with six-week sprints, for example, the task force can drive swift, data-driven decision-making.

Continuous improvement in the transformation will remain paramount, with regular evaluation of Work3 initiatives and refinement of strategies based on feedback and evolving organizational needs. Crucially, the task force ensures optimal talent allocation, matching the right people to the right projects at the right time.

Identify key stakeholders

- Map out a diverse range of stakeholders, including C-suite executives, department heads, HR leaders, frontline employees, and external consultants.
- Aim for a representative sample of at least 10 percent of your workforce across all levels and departments.

Define objectives and protocols

- Establish clear, measurable objectives for the diagnostic process, such as "Identify top three barriers to AI adoption" or "Assess current blockchain literacy across departments."
- Develop a structured interview protocol with a mix of quantitative and qualitative questions.
- Include AI-readiness assessments, web3 knowledge checks, and cultural adaptability metrics.

Conduct strategic interviews

- Select participants strategically, ensuring representation across departments, tenure, and job functions.
- Employ active listening techniques and "yes . . . and" language to encourage open dialogue.
- Use AI-assisted transcription and analysis tools to identify patterns and themes in real time.

Ask these sample questions, which we have used at the Work3 Institute:

- How familiar are you with AI and web3 technologies (such as the metaverse, digital twins, and so on) currently being used or in the marketplace?
- What specific AI tools or applications do you think could most benefit your organization in the near future?
- What skills or training do you think employees will need to effectively work alongside AI systems?
- What potential resistance to AI adoption do you foresee within the organization, and how might you address it?

- What opportunities do you see for AI to improve customer experiences or create new products or services?
- How might AI affect your data management and cybersecurity practices?
- How can you ensure your AI and web3 adoption aligns with your company's core values and mission?

Analyze and synthesize findings

- Utilize machine learning algorithms to identify correlations and insights across interviews.
- Create a comprehensive "Work3 Readiness Index" based on aggregated data.
- Develop visual dashboards to present findings to leadership in an accessible format.

Share the results of your findings with everyone—key stakeholders, leadership teams, and employees at large. Encourage dialogue, feedback, and collaboration to ensure that everyone feels heard and invested in the transformation journey. Use the insights gathered to build a compelling case for change and garner buy-in for the Work3 transition.

Step 3: Develop transparent communication channels

Develop a decentralized communication plan that outlines key messages, communication channels, and targeted audiences to ensure that everyone within the organization is well-informed and knows whom to turn for questions and necessary resources. It is critical to see your workers as partners, so encouraging two-way communication is a must.

One way you can offer transparent communication channels is to implement robust feedback mechanisms and interactive Q&A

sessions to create an open dialogue between leadership and employees. Host regular virtual town halls using immersive technologies like virtual reality to create a more engaging and personal experience, even in remote settings. Utilize blockchain-based voting systems for key decisions, ensuring every voice is heard and recorded transparently. Consider implementing an AI-powered suggestion platform where employees can submit ideas anonymously, with machine learning algorithms categorizing and prioritizing submissions for review.

Another way to develop transparent communication channels is to create dedicated Slack channels or use decentralized communication platforms for ongoing discussions about the transformation. Implement a gamified system that rewards employees for actively participating in these channels, contributing ideas, and helping colleagues understand new Work3 concepts. This not only encourages engagement but also accelerates the dissemination of knowledge throughout the organization.

After you have a successful two-way communication mechanism in play, regularly conduct pulse surveys using advanced analytics to gauge employee sentiment and identify emerging concerns in real time. Use natural language processing (NLP), a machine learning technology that gives computers the ability to interpret, manipulate, and comprehend human language to analyze open-ended responses, providing deeper insights into employee perspectives. Share aggregated results openly, demonstrating a commitment to transparency and showing how employee feedback directly influences the transformation strategy.

By creating these multifaceted, technology-enabled communication channels, you'll not only keep employees informed but also actively involve them in shaping the Work3 journey. This approach fosters a cohesive digital work environment that balances synchronous engagement with asynchronous productivity, optimizing collaboration across time zones and geographical boundaries.

Step 4: Forge a cutting-edge technological foundation

This critical step demands strategic investments in AI and other disruptive technologies to propel your organization into the Work3 era. The transition to Work3 is not merely an upgrade of existing systems; it's a fundamental reimagining of your technological infrastructure. The Work3 technological ecosystem is characterized by its interconnectedness, adaptability, and intelligence. It's a symbiosis of human creativity and machine efficiency, where AI augments human decision-making, blockchain ensures trust and transparency, and immersive technologies break down physical barriers.

As you embark on this transformation, it's crucial to approach it holistically. You should view each technological component not in isolation but as part of an integrated system that drives your organization's Work3 vision. This requires a paradigm shift in how you think about technology—moving from siloed solutions to an interconnected, intelligent infrastructure that permeates every aspect of your operations.

Focus on four key areas:

AI integration. Deploy state-of-the-art AI algorithms across all business functions. Implement NLP for enhanced communication, machine learning for predictive analytics, and computer vision for quality control. Utilize platforms like TensorFlow or PyTorch to develop custom AI solutions tailored to your specific needs.

Cloud infrastructure. Move your organization's digital operations to a flexible and powerful cloud-based system. Use multiple cloud service providers to keep your options open and avoid becoming too dependent on any single company. Package your software in a way that makes it easy to move and manage across different systems, ensuring your technology can grow smoothly with your business needs.

Web3 technologies. Adopt new internet technologies that enhance data security and give users more control. Use digital systems that automatically execute agreements and store information across multiple computers instead of in one central location. Create digital rewards that encourage participation and loyalty from employees, customers, and partners. These technologies can make your organization more transparent, secure, and engaging.

Immersive technologies. Invest in cutting-edge AR/VR solutions. Utilize platforms like Unity or Unreal Engine to create immersive training simulations. Implement Microsoft HoloLens or Oculus for Business to facilitate remote collaboration and virtual product prototyping.

To execute this technological transformation:

- Form strategic partnerships with leading tech vendors and AI research institutions.
- Recruit top-tier AI engineers, blockchain developers, and cloud architects.
- Establish an intensive upskilling program for your existing workforce, focusing on data science, blockchain, and cloud computing.
- Create a dedicated innovation lab to continuously explore and test emerging technologies.

While this overview merely scratches the surface of establishing a robust technological foundation for the Work3 era, its implications are profound. By strategically and boldly embracing these cutting-edge advancements, your organization isn't just keeping pace; it's positioning itself at the vanguard of industry evolution.

This technological metamorphosis catalyzes a new level of organizational dynamism. It fosters unprecedented agility, allowing your enterprise to pivot swiftly in response to market shifts. It amplifies your innovation capacity, transforming every team member into a potential disruptor. In essence, this isn't merely about adopting new tools but about designing a living, breathing technological ecosystem that evolves in symbiosis with your business objectives.

Step 5: Design an agile, decentralized organizational matrix

This pivotal step involves a comprehensive reimagining of organizational structures to align with the fluid, interconnected nature of the Work3 ecosystem. The transition to a decentralized organizational matrix represents a paradigm shift in how we conceive of and operate businesses in the digital age. It's not merely about flattening hierarchies; it is a new paradigm where power is distributed, decision-making is democratized, and innovation can spring from any node in the network.

The goal is to create an organization that's as nimble and responsive as a startup, yet able to leverage the resources and reach of a large enterprise. This requires a fundamental rethinking of the traditional concepts of leadership, accountability, and collaboration. The decentralized organizational matrix is designed to harness the collective intelligence of your workforce, breaking down silos and fostering cross-pollination of ideas. It recognizes that in a world of rapid technological change, the best ideas can come from anywhere in the organization. By empowering employees at all levels to contribute and innovate, you create a more engaged, motivated workforce and a more resilient organization.

However, implementing such a structure is not without challenges. It requires a cultural shift, new skills, and robust technological support. Leaders must learn to influence without direct control, employees must become comfortable with increased autonomy and responsibility, and systems must be put in place to ensure coordination and alignment across the decentralized structure.

Consider implementing the following strategies:

Holacracy-inspired restructuring. Replace traditional hierarchies with a dynamic circle structure. Each circle operates autonomously while interconnecting with others, fostering rapid decision-making and adaptability. Utilize platforms like HolacracyOne or Holaspirit to manage this transition.

OKR-driven performance management. Implement objectives and key results (OKRs) at all levels. Use tools like Lattice or 15Five to cascade and align goals, ensuring transparency and accountability across the organization.

AI-powered team formation. Utilize AI algorithms to dynamically assemble cross-functional teams based on project requirements, skills, and past collaboration data. Explore platforms like Asana Teammates, Pygmalion AI, or Orgnostic for intelligent team composition.

Decentralized knowledge management. Deploy a blockchain-based knowledge-sharing system using platforms like Steem or Hive to incentivize content creation and curation across the organization.

By flattening the organization and adopting a holacracy-inspired restructuring, this approach not only enhances operational efficiency but also cultivates a culture of autonomy, ownership, and continuous improvement.

Step 6: Pilot a DAO within core teams

As we discuss in chapter 6, a DAO is a revolutionary organizational model that leverages blockchain technology to create a transparent, democratic decision-making Work3 ecosystem. Unlike traditional hierarchies, DAOs distribute power among members and create an organization that's as nimble and responsive as a startup, yet able to leverage the resources and reach of a large enterprise.

DAOs recognize that the best ideas can come from anywhere in the organization. By empowering employees at all levels to contribute and innovate, you create a more engaged, motivated workforce and a more resilient organization. It does, however, require a cultural shift, new skills, and robust technological support. Leaders must learn to influence without direct control, employees must become comfortable with increased autonomy and responsibility, and systems must be put in place to ensure coordination and alignment across the decentralized structure.

Here's how to implement a DAO pilot:

- *Define clear objectives.* Establish specific, measurable goals for the DAO pilot, such as "Reduce decision-making time by 40 percent and increase cross-departmental collaboration by 50 percent within three months."
- *Select pilot scope.* Choose a strategic cross-functional team of ten to fifteen members working on a high-impact project. This focused approach allows for easier management and evaluation of the DAO's effectiveness.
- *Deploy user-friendly technology.* Implement an intuitive DAO-creation platform like Aragon or DAOstack. These tools provide a low-barrier entry point for setting up governance structures without requiring deep technical expertise.
- *Establish governance protocol.* Develop clear, transparent guidelines for decision-making processes; for instance, "Any team member can propose an initiative. Proposals achieving 60 percent approval with at least 75 percent participation rate will be implemented."
- *Implement token economics.* Design a token distribution system that aligns with your organizational values and objectives; for example, allocate a hundred governance tokens to each member initially, with the potential to earn additional tokens based on contributions and performance.
- *Set up efficient voting mechanisms.* Utilize a secure, user-friendly voting platform such as Snapshot for cost-effective and accessible decision-making. Ensure the system is easily accessible via various devices to maximize participation.

- *Provide comprehensive training.* Conduct in-depth workshops explaining DAO concepts, platform usage, and new decision-making processes. Offer ongoing support through dedicated communication channels and a help desk.
- *Implement robust analytics.* Leverage blockchain-native analytics tools like Dune Analytics to create real-time dashboards for monitoring key performance indicators and participation metrics.

Based on pilot results and feedback, refine the DAO structure and governance model. If successful, develop a phased approach for expanding the DAO system to other departments or projects within the organization. By implementing this DAO pilot, you're embracing a more democratic, transparent, and agile way of working. This approach can lead to faster decision-making and significant increased employee engagement.

Step 7: Unveil the Work3 transformation plan

We purposely wait until step seven to unveil the Work3 transformation plan. We want technologies to be integrated and piloted so that by the time you unveil the plan, your workforce is on the train to begin the journey. The plan, which the CEO should unveil at a town-hall-style meeting, paints a vivid picture of the transformed workplace, highlighting the benefits and opportunities that Work3 will bring to individuals, teams, and the organization as a whole. The plan should address not only the "what" and "how" of the transformation but also the crucial "why," helping employees understand the imperative for change and their role in shaping the organization's future.

Moreover, this unveiling should be seen as the launch of an ongoing dialogue rather than a onetime event. It marks the beginning of a sustained communication campaign that will continue throughout the transformation journey, keeping employees informed, engaged, and empowered. The strategy should leverage a mix of traditional and cutting-edge communication tools, from town halls and newsletters to immersive VR experiences and AI-powered personalized updates.

Here's our suggested detailed approach:

Develop a multichannel communication plan

- Host a companywide virtual town hall
- Produce a series of engaging video explainers
- Launch a weekly "Transformation Tuesday" email newsletter

Articulate clear transformation objectives

- Set Specific, Measurable, Achievable, Relevant, and Time-bound (SMART) goals; for example, "Increase remote work productivity by 30 percent within twelve months"
- Outline specific KPIs for each department

Present a detailed timeline

- Break down the transformation into quarterly milestones
- Use visual tools like Gantt charts to illustrate the progression

Highlight technological innovations

- Showcase AI integration plans; for example, "Implementing GPT-4 for customer service by Q3"
- Explain blockchain adoption strategy; for example, "Piloting supply chain traceability in Q4"

Address workforce impact

- Outline reskilling initiatives; for example, "Launching 'AI Academy' with fifty courses by next month"
- Discuss new roles creation; for example, "Hiring twenty blockchain developers over the next six months"

Create a change ambassador program

- Recruit fifty employees across departments as change champions
- Provide them with advanced training and exclusive transformation insights

Launch a "Day in the Life of Work3" campaign

- Produce immersive VR experiences showcasing future work scenarios
- Host "Future Friday" events where teams can try out upcoming technologies

By bringing in your workers as partners in the Work3 transition, you inform and also inspire and engage your workforce in the Work3 transformation journey. This approach ensures transparency, builds excitement, and creates a shared vision for the future of work in your organization.

Step 8: Implement comprehensive Work3 boot camps

The Work3 boot camps are intensive, immersive learning experiences that blend rigorous training with long-term reinforcement. Each boot camp spans two weeks of concentrated, hands-on learning, followed by three months of micro-learning sessions to ensure knowledge retention and practical application. We've adopted a hybrid model, seamlessly integrating in-person workshops with virtual learning modules to cater to diverse learning styles and maximize engagement. The boot camps are conducted quarterly, with the ambitious goal of cycling all employees through the program within an eighteen-month time frame. To foster optimal interaction and personalized attention, each cohort is carefully curated to include twenty-five to thirty participants, striking a balance between diverse perspectives and intimate group dynamics. This structure ensures a comprehensive yet tailored learning journey that equips employees with the skills and mindset necessary to thrive in the Work3 era.

A suggested curriculum is:

AI fundamentals (two days)

- Machine learning and deep learning basics
- Neural networks and their applications
- Hands-on exercises using platforms like TensorFlow or PyTorch
- Case studies: AI implementation across industries

Web3 essentials (two days)

- Blockchain technology and consensus mechanisms
- Smart contracts and decentralized applications (dApps)
- Tokenomics and decentralized finance (DeFi) principles
- Practical workshop: creating a simple smart contract

Practical applications (two days)

- AI use cases: predictive analytics, NLP, computer vision
- Web3 applications: supply chain traceability, decentralized identity, NFTs in business
- Group project: developing a Work3 solution for a real business challenge
- Presentation and peer review of projects

Ethical considerations and governance (one day)

- AI ethics: bias, transparency, and accountability
- Web3 governance models and digital sovereignty

- Data privacy and security in the Work3 era
- Panel discussion with industry experts on responsible tech adoption

Work3 culture and leadership (one day)

- Leading in a decentralized environment
- Fostering innovation and continuous learning
- Change management in the Work3 transition
- Personal action planning for Work3 implementation

The Work3 boot camp experience extends far beyond the initial intensive training period, evolving into a comprehensive, ongoing learning journey. Post-boot-camp engagement is meticulously designed to reinforce and expand on key concepts through weekly micro-learning modules, ensuring continuous skill development. Monthly virtual meetups serve as a platform for boot camp alumni to share experiences and insights, fostering a community of practice. Quarterly hackathons provide hands-on opportunities to apply Work3 principles to real business challenges, bridging the gap between theory and practice.

To gauge the effectiveness of this approach, we implement a rigorous measurement and evaluation process. This includes pre- and post-boot-camp assessments to quantify knowledge gain; thirty-, sixty-, and ninety-day follow-ups to track the application of learned skills in the workplace; and a comprehensive impact analysis that monitors Work3 initiatives launched by boot camp participants. This multifaceted approach ensures the retention and application of crucial Work3 concepts and provides valuable data to continually refine and optimize the boot camp experience for maximum organizational impact.

Step 9: Commit to a dynamic continuous learning ecosystem

To thrive in the Work3 era, organizations must cultivate a culture of perpetual learning and adaptation. This step involves creating a comprehensive, multifaceted approach to learning that permeates every aspect of the organization. You can begin by implementing an AI-powered learning management system (LMS) like Degreed or EdCast XP.

A dynamic continuous learning ecosystem should personalize learning paths for each employee based on their role, skills, and career aspirations, with a target of forty learning hours per employee per quarter. Complement this with a "Future Skills Academy" offering courses on emerging technologies and partnering with top-tier universities for micro-credentials. Set an ambitious goal: 50 percent of employees certified in at least one future skill within eighteen months.

To foster innovation and cross-pollination of ideas, establish several key initiatives:

- *Learning sabbaticals.* Allow 5 percent of your workforce annually to take one- to three-month paid leaves for intensive learning experiences.
- *Cross-functional innovation squads.* Form diverse teams to solve real business challenges using design thinking, rotating 20 percent of employees through these squads each year.
- *Reverse mentoring program.* Pair junior employees with senior leaders to share insights on emerging trends, aiming for 100 percent executive participation.

Embrace failure as a learning opportunity by implementing a "fail forward" framework. Host quarterly "failure festivals" to destigmatize and learn from mistakes, aiming to capture and analyze insights from 100 percent of failed projects. This approach cultivates resilience and encourages calculated risk-taking.

To democratize knowledge sharing, launch an internal TED Talks series and a Skill Marketplace. Target fifty employee-led talks per year with 80 percent attendance and facilitate a thousand peer-to-peer skill exchanges in the first year. These initiatives empower employees to become both teachers and learners.

Finally, make learning a core business metric. Establish "Learning KPIs" for leadership, tying 20 percent of leadership bonuses to team learning metrics. Create an "Emerging Tech Lab" where a hundred employees per quarter can have immersive experiences with cutting-edge technologies.

Step 10: Uphold the ten operating principles of Work3

The ten operating principles of Work3 that we cover in chapter 2 (and summarize in the sidebar here) are the governing tenets that drive an organization forward into a Work3 ecosystem. By following these principles, you will provide an entirely new model of work in which contributors can feel greater ownership of their work, greater flexibility in how they work, and expanded agency to work from wherever they are. Employees gain the freedom to design their workdays in ways that best suit their productivity peaks and personal commitments, which cultivates a culture of trust and respect.

By integrating these ten operating principles into your organization's DNA, you're not just changing policies; you're igniting a revolution in how work is conceived and executed. Picture a workplace where blockchain developers collaborate seamlessly with marketing creatives, where AI-powered analytics inform strategy in real time, and where even the most junior team member can propose ideas that reshape entire business units.

This isn't utopian thinking; it's the new reality of Work3. Your employees won't just feel valued; they'll be empowered by smart contracts that automatically recognize and reward their contributions. They'll navigate a tokenized ecosystem where their skills and innovations directly impact their stake in the company's success.

Imagine town halls conducted in virtual reality, where geographically dispersed teams gather in immersive digital spaces to cocreate the company's future. Or consider how decentralized decision-making protocols could transform traditional hierarchies, allowing for fluid, project-based structures that adapt in real time to market changes.

The Ten Operating Principles of Work3

Partnership. View work as a collaborative partnership to foster a sense of shared purpose and drive collective success.

Transparency. Communicate openly and honestly in the workplace, ensuring everyone has access to information and understands what's happening.

Autonomy. Trust employees to make decisions independently in order to breed innovation and productivity.

Ownership. Offer equity to workers and help them to see their job as part of a larger endeavor.

Decision-making. Empower individuals to make important decisions swiftly based on their expertise, enhancing agility and efficiency within the organization.

Flexibility. Allow employees to work according to their own schedule and in their preferred environment.

Upskilling. Stay relevant in a rapidly evolving landscape by identifying skill gaps and seeking out learning opportunities to ensure competitiveness and growth.

Incentives. Recognize and reward work that aligns with organizational goals and reinforces their value within the organization.

Interoperability. Ensure seamless collaboration across different tools and systems.

Community. Cultivate a supportive and inclusive community within the workplace.

Each principal acts as a force multiplier, creating a workplace that's not just satisfying but exhilarating.

Step 11: Implement advanced people analytics

In the rapidly evolving Work3 landscape, traditional HR metrics are no longer sufficient. Advanced people analytics harnesses the power of

AI and big data to provide deep insights into your workforce, enabling data-driven decision-making and strategic workforce planning.

People analytics goes beyond basic HR data, integrating information from various sources to create a holistic view of your organization's human capital. It allows leaders to understand complex patterns in employee behavior, predict future trends, and make proactive decisions to optimize workforce performance and engagement.

We suggest implementing a cutting-edge people analytics system this way:

AI-powered talent intelligence platform. Implement advanced AI systems like Eightfold AI or Pymetrics to analyze employee data comprehensively. These platforms use machine learning algorithms to predict skill gaps, identify retention risks, and match employees to optimal roles. By leveraging these insights, aim to reduce the time to hire by 30 percent and improve retention by 25 percent within the first year.

Real-time engagement monitoring. Deploy tools such as Perceptyx or Qualtrics for continuous pulse surveys. These platforms use NLP to analyze open-ended feedback, providing nuanced insights into employee sentiment. Set a goal for weekly micro-surveys with an 80 percent response rate to maintain a real-time understanding of your workforce's engagement levels.

Network analysis for collaboration insights. Implement organizational network analysis tools like TrustSphere to map informal networks within your organization. This analysis reveals key influencers and collaboration patterns in the Work3 environment, allowing you to optimize team structures and communication flows. Target a 50 percent increase in cross-functional collaboration within six months.

This data-driven approach enables leaders to make informed decisions that enhance employee experience, optimize talent utilization, and drive Work3 transformation success.

Step 12: Establish a transformation governance framework

Current traditional governance models are often too rigid to keep up with the pace of change. A dynamic transformation governance framework provides the structure needed to guide the transformation while maintaining the flexibility to adapt quickly to new challenges and opportunities.

This approach combines agile methodologies with cutting-edge technologies to create a responsive, data-driven governance system. It ensures that your transformation efforts remain aligned with strategic objectives while fostering a culture of continuous improvement and innovation.

Establish this framework by implementing the following elements of the structure:

Agile steering committee. Implement biweekly sprint reviews and monthly steering meetings to maintain momentum. Utilize digital kanban boards (e.g., Jira) for real-time progress tracking, allowing for swift identification and resolution of bottlenecks. Set OKRs for each transformation work stream, reviewing them quarterly to ensure alignment with overall goals.

AI-powered decision support system. Deploy an AI-driven scenario planning tool like Palantir Foundry to enhance decision-making. These systems can process vast amounts of data to generate insights and predictions, enabling more informed strategic choices. Conduct monthly simulations to test potential transformation pivots, aiming for 80 percent of strategic decisions to be data-informed within six months.

Real-time transformation dashboard. Implement a live dashboard using tools like Tableau or Power BI, accessible to all employees. This transparency fosters engagement and accountability across the organization. Track key metrics such as AI adoption rates, blockchain transactions, and skill acquisition. Set a goal for 100 percent of employees to interact with the dashboard monthly, ensuring widespread understanding of transformation progress.

By implementing this dynamic governance framework, organizations can ensure their Work3 transformation remains agile, data-driven, and aligned with both internal goals and external best practices.

. . .

The issue of a timeline for this transformation is not about reaching a final destination but about embarking on an ongoing journey of evolution. The Work3 transformation is a continuous process, adapting to the ever-changing technological landscape and market dynamics. However, what we can definitively say is this: our twelve-step Work3 transformational road map is a powerful catalyst for organizational metamorphosis. It's not just a framework but a launchpad for exponential growth and innovation. Each step is designed to create a compounding effect, where progress in one area accelerates advancements in others.

This road map doesn't just monitor progress; it propels it. It transforms problem-solving from a reactive exercise to a proactive, AI-assisted strategy. Decision-making evolves from hierarchical pronouncements to blockchain-enabled consensus building. Stakeholder engagement shifts from periodic check-ins to continuous, immersive collaboration in virtual and augmented realities.

By amplifying collaboration through DAOs, ensuring transparency via distributed ledger technologies, and fostering agility through AI-powered predictive analytics, the road map doesn't just prepare your organization for the future; it empowers you to shape it. It's a blueprint for turning challenges into opportunities and for steering your organization toward a future where innovation isn't just encouraged; it's inevitable. In essence, while the transformation is ongoing, the impact of implementing this road map is immediate, tangible, and exponentially rewarding. It's not about how long the journey takes but about how quickly you can start reaping the benefits of being at the forefront of the Work3 revolution.

Considerations for People-Focused Leaders

In a sea of technological advancements, it is easy to get lost in the technology and forget about a key element of any organization: the people. We are often asked how organizations can shift their culture to be more focused on people: "How can we communicate to our workers that we care? How can we walk the walk and talk the talk?"

The answers lie in intentional actions that prioritize the well-being and empowerment of employees. It starts by cultivating open channels of communication, where employees feel heard and valued. People-focused, transparent leadership that shares openly and involves employees in decision-making processes builds trust and engenders a sense of ownership. Investing in employee development and well-being initiatives demonstrates a commitment to the holistic growth and fulfillment of workers. This includes providing opportunities for skill development, promoting work-life balance, and offering support for mental health and wellness.

Furthermore, creating a culture of inclusivity and diversity ensures that all voices are heard and valued within the organization. By embracing diverse perspectives and experiences, companies can harness the full potential of their workforce and drive innovation and creativity. Here are some additional considerations for people-focused leaders.

The Work3 paradigm represents a quantum leap in how we approach labor, time, and societal impact. By leveraging AI, blockchain, and decentralized systems, we're optimizing workflows and liberating human potential. This revolution promises to reclaim countless hours lost to inefficiency, creating space for creativity, innovation, and personal fulfillment. Moreover, Work3 acts as a great equalizer, dismantling traditional barriers to entry and democratizing opportunities across global talent pools. Through blockchain-verified credentials and AI-powered skill matching, merit truly becomes the currency of success, regardless of background or location.

Crucially, this new model transcends the myopic view of work as a siloed activity. It interweaves labor with broader societal and environmental concerns, making every task an opportunity for positive impact. Smart contracts can automatically allocate resources to carbon offset programs, while DAOs can pivot entire industries toward sustainable practices with unprecedented agility.

Appoint a chief heart officer

Have a chief heart officer (CHO) lead in designing a better employee experience. Imagine a workplace where happiness isn't just a byproduct but a top priority—a place where employees feel valued, supported, and truly happy to come to work every day. As the mastermind behind employee experience, the CHO dreams up creative initiatives to bring teams closer together, from team-building activities to inspiring recognition programs that celebrate big and small wins.

Communication is key, and the CHO is responsible for creating open, honest channels for feedback and dialogue, ensuring that every voice is heard and valued. Whether they're resolving conflicts, addressing concerns, or simply lending an empathetic ear, they're there to support and uplift the entire team. But perhaps most important, the CHO is a catalyst for growth and development. They believe in the power of lifelong learning and are committed to helping employees reach their full potential. With personalized training, mentorship programs, and opportunities for advancement, they empower individuals to carve out their own paths to success.

In the end, CHO isn't just a job title; it's a philosophy, a mindset, and a promise. The CHO is a wellness warrior, dedicated to nurturing the physical, mental, and emotional well-being of every team member. From yoga classes to mindfulness workshops, the CHO is on a mission to help employees find balance, recharge, and thrive in both their personal and professional lives. It's about creating a workplace where happiness isn't just a perk; it's the foundation of everything you do.

Commit to flexible work policies

This strategy recognizes the transformative power of aligning work with individual circadian rhythms and life commitments. It's not merely about flexibility but about optimizing human potential by tapping into each person's unique productivity peaks.

Consider the software engineer who enters a state of flow at 2 a.m., crafting elegant code in the quiet hours, or the parent who structures their workday around their children's schedules, seamlessly blending professional output with family engagement. This approach doesn't just accommodate these patterns, but celebrates them as catalysts for innovation and job satisfaction.

At the Work3 Institute, we've pioneered initiatives that elevate this concept:

- *"Mindfulness Fridays."* Dedicated time for deep, uninterrupted focus, free from the tyranny of back-to-back meetings
- *Quarterly "heart checks."* A blockchain-enabled, anonymous feedback system allowing team members to candidly assess workload, compensation, and overall job satisfaction
- *AI-powered scheduling optimization.* Leveraging machine learning to identify and suggest ideal work windows for each employee based on their past performance data and stated preferences
- *"Emoji economics."* Encouraging the use of digital shorthand and informal language in corporate communications to foster authenticity and rapid idea exchange

These initiatives aren't just feel-good policies but strategic imperatives. By aligning work patterns with individual biorhythms and life circumstances, organizations unlock unprecedented levels of creativity, efficiency, and employee loyalty.

Implement a DAO structure to manage these flexible arrangements, allowing team members to collectively evolve policies in real time. Use smart contracts to automatically adjust workloads and compensation based on output rather than hours logged.

The result? A workforce that's fully engaged, bringing their whole selves to their roles and driving innovation at every level. This isn't just about being flexible but about fundamentally reimagining the relationship between work and human potential in the Work3 era. This strategy is one of our favorites because it recognizes the importance of flexible arrangements that allow people to work when they are at their best—tapping into their rhythms. For example, many engineers like to plug in very late at night to do their work, and others like to take time out during the day to run or go to their kids' soccer practices. The key is to allow people to work when they want, not according to rigid and arbitrary nine-to-five schedules.

Embrace a distributed and global workforce

In the Work3 era, geographical boundaries are becoming increasingly irrelevant. Forward-thinking companies are adapting to remote work and leveraging it as a strategic advantage to build diverse, high-performing teams that span the globe.

Remote-first organizations are outpacing their competition by tapping into a global talent pool, offering the flexibility that top performers demand. This is about accommodating employee preferences and fundamentally reimagining how work is structured and executed.

Consider implementing a "follow the sun" model, where core collaboration hours overlap across time zones, complemented by asynchronous work periods. This approach maximizes productivity and ensures 24/7 operational capability. Utilize AI-powered scheduling tools to optimize these overlaps, ensuring seamless global collaboration.

To foster a sense of community in this distributed landscape, go beyond traditional virtual check-ins. Implement immersive VR team-building experiences, blockchain-based reward systems for cross-cultural collaboration, and AI-facilitated random coffee meetups to simulate serendipitous office interactions.

Performance evaluation in this new paradigm must evolve. Implement blockchain-verified, outcome-based assessment systems that track deliverables and impact, not hours logged. Utilize machine learning algorithms to identify patterns of high performance and replicate them across teams.

Moreover, consider adopting a DAO structure for certain projects or departments. This allows for truly democratic decision-making and resource allocation, regardless of physical location or time zone.

By embracing these innovative approaches, HR leaders can design a workforce that's distributed and dynamically interconnected—a global neural network of talent that's constantly learning, adapting, and innovating. This isn't just about remote work but about creating a borderless, always-on ecosystem of human potential that can respond to market changes with unprecedented agility and insight.

Adopt societal considerations

In essence, Work3 isn't just redefining our jobs but is reimagining our role as global citizens, harmonizing professional aspirations with planetary stewardship. It's a paradigm where economic growth and societal progress are compatible and inextricably linked, ushering in an era of work that's not only more efficient and fairer but profoundly more meaningful.

Accenture's Net Better Off framework is a business strategy and approach that focuses on creating value for the company as well as all stakeholders involved, including employees, customers, suppliers, communities, and the environment (see the sidebar "Elements of Accenture's Net Better Off Framework"). The framework aims to ensure that the overall impact of the company's actions and decisions results in a net positive outcome for society as a whole.

At its core, the framework emphasizes the importance of considering the broader social and environmental implications of business activities, beyond just financial metrics. It encourages companies to adopt a holistic approach to value creation that considers the long-term sustainability and well-being of all stakeholders.

By embracing the Net Better Off framework, companies will enhance their reputation and brand value and drive long-term business success by creating shared value for all stakeholders and contributing to a more sustainable and equitable future.

Elements of Accenture's Net Better Off Framework

Stakeholder engagement. Actively engage with stakeholders to understand their needs, concerns, and expectations, and incorporate their perspectives into decision-making processes.

Impact measurement. Develop metrics and indicators to assess the social, environmental, and economic impact of business activities, and use this information to guide strategic planning and performance evaluation.

Sustainable practices. Integrate sustainability principles into business operations, such as reducing carbon emissions, minimizing waste, and promoting ethical sourcing and labor practices.

Inclusive growth. Foster inclusive growth by creating opportunities for marginalized and underserved communities, promoting diversity and inclusion within the workforce, and supporting local economic development initiatives.

Innovation. Encourage innovation and collaboration to develop solutions that address societal challenges and contribute to positive social and environmental outcomes.

The Dawn of Work3

As we conclude our exploration of the Work3 transformation road map and this book, we find ourselves at a pivotal moment in human history. The convergence of AI and web3 technologies are reshaping our workplaces and redefining the very essence of what it means to work, create, and contribute to society. This revolution promises a future where the boundaries between the physical and digital worlds blur, where our potential is amplified by intelligent machines, and where the power to shape our professional destinies rests firmly in our own hands.

We can start to imagine a world where our workplace is no longer confined to a physical space but exists in a rich, immersive digital realm. Here, you collaborate with colleagues from across the globe in virtual environments that feel as real as any office. AI assistants anticipate your needs, prioritize your priorities, and do a lot of the work for you, enhancing your creativity and productivity in ways we can only begin to fathom. Smart contracts ensure that your contributions are fairly rewarded, while DAOs give you a true voice in the direction of your company. In this future, your skills and ideas are your most valuable currency, verified on the blockchain and recognized across a global network of opportunities.

But this technological revolution is even more than the transformation itself; it's about democratizing opportunity, giving us back time to pursue human interests outside of work, and reclaiming our humanity in the world of work. As AI takes on routine tasks, we are freed to focus on what makes us uniquely human: our creativity, empathy, and ability to solve complex problems. The Work3 era promises a more equitable world, where opportunities are not limited by geography or socioeconomic backgrounds but are open to all who wish to learn and contribute. It's a world where work and life are not at odds, but in accord, allowing us to pursue our passions and purpose with unprecedented freedom and flexibility.

Yet, as we embrace this exciting future, we must remain mindful of the challenges it presents. The rapid pace of change will require us to become lifelong learners, constantly adapting to new technologies and ways of working. We must vigilantly guard against the potential misuse of AI and ensure that the decentralized web remains a force for empowerment, not exploitation. As leaders, we have a profound responsibility to guide this transformation with wisdom, empathy, and a commitment to the greater good.

The journey to Work3 is not just a technological evolution; it's a revolution in how we perceive work, value, and human potential. It calls us to reimagine our organizations, economies, and societies. Let us embrace this future with open minds and hearts. The power to shape this new world of work lies within each of us. Together, we can create a future where work is not just a means of survival but a path to fulfillment, growth, and collective progress. The era of Work3 is dawning. Let us step boldly into this new day, ready to build a more empowering, equitable, and human-centric world of work.

NOTES

Introduction

1. Ryan Pendell, "Employee Engagement Strategies: Fixing the World's $8.8 Trillion Problem," Gallup, June 14, 2022, https://www.gallup.com/workplace/393497/world-trillion-workplace-problem.aspx.

2. GitLab, "GitLab Culture," GitLab Handbook, n.d., https://handbook.gitlab.com/handbook/company/culture/.

3. Eightfold.ai, "Everything Talent, Powered by AI," n.d., https://eightfold.ai/.

4. The Sandbox, "An Impetus for Metaverse," The Sandbox Documentation, n.d., https://docs.sandbox.game/en/what-is-tsb/metaverse-jobs.

5. Work3 Institute, "Human-Centric Digital Transformation," n.d., https://work3.me/.

Chapter 1

1. HCA Healthcare, "It's Official! 'Burnout' Defined and Recognized as a Medical Diagnosis," *HCA Healthcare Today*, June 11, 2019, https://hcahealthcaretoday.com/2019/06/11/its-official-burnout-defined-and-recognized-as-a-medical-diagnosis/#:~:text=If%20so%2C%20you%20may%20be,medical%20providers%20in%20diagnosing%20diseases.

2. Jim Harter, "In New Workplace, U.S. Employee Engagement Stagnates," Gallup, January 23, 2024, https://www.gallup.com/workplace/608675/new-workplace-employee-engagement-stagnates.aspx#:~:text=For%20the%20full%20year%20of,June%20of%20the%20same%20year.

3. Elizabeth J. Altman et al., *Workforce Ecosystems: Reaching Strategic Goals with People, Partners, and Technologies* (Cambridge, MA: MIT Press, 2023), https://direct.mit.edu/books/monograph/5563/Workforce-EcosystemsReaching-Strategic-Goals-with.

4. David De Cremer and Garry Kasparov, "AI Should Augment Human Intelligence, Not Replace It," hbr.org, March 18, 2021, https://hbr.org/2021/03/ai-should-augment-human-intelligence-not-replace-it.

5. United Nations, "Tools Like ChatGPT Likely to Complement Jobs, Not Destroy Them: ILO," *UN News*, August 22, 2023, https://news.un.org/en/story/2023/08/1139972.

6. Sheila Chiang, "IBM CEO Says AI Will Impact White-Collar Jobs First, but Could Help Workers Instead of Replacing Them," CNBC, August 22, 2023, https://www.cnbc.com/2023/08/22/ibm-ceo-says-ai-will-impact-white-collar-jobs-first.html.

7. Touchcast, June 7, 2024, https://touchcast.com/.

8. Goldman Sachs, "Generative AI Could Raise Global GDP by 7%," Goldman Sachs, April 5, 2023, https://www.goldmansachs.com/intelligence/pages/generative-ai-could-raise-global-gdp-by-7-percent.html.

9. World Economic Forum, "Future of Jobs Report 2023: Up to a Quarter of Jobs Expected to Change in Next Five Years," press release, May 1, 2023, https://www.weforum.org/press/2023/04/future-of-jobs-report-2023-up-to-a-quarter-of-jobs-expected-to-change-in-next-five-years/.

10. Goldman Sachs, "Generative AI Could Raise Global GDP by 7%."

11. Goldman Sachs, "AI Investment Forecast to Approach $200 Billion Globally by 2025," August 1, 2023, https://www.goldmansachs.com/intelligence/pages/ai-investment-forecast-to-approach-200-billion-globally-by-2025.html.

12. Steve Glaveski, "How DAOs Could Change the Way We Work," hbr.org, April 7, 2022, https://hbr.org/2022/04/how-daos-could-change-the-way-we-work.

13. Hilary Weaver, "How a 52-Year-Old Word Invented by a *Vogue* Editor Became 2017's Word of the Year," *Vogue*, December 15, 2017, https://www.vanityfair.com/style/2017/12/youthquake-is-oxford-dictionary-word-of-the-year.

14. US Census, "U.S. and World Population Clock," https://www.census.gov/popclock/.

15. Roblox, Q1 2023 Supplemental Materials, May 10, 2023, https://s27.q4cdn.com/984876518/files/doc_financials/2023/q1/Q1-23-Supplemental-Materials-FINAL.pdf.

16. Jason Wise, "How Many Kids Want to Be YouTubers?," EarthWeb, October 24, 2023, https://earthweb.com/why-children-want-to-be-youtubers-when-they-grow-up/.

17. Goldman Sachs, "The Creator Economy Could Approach Half-a-Trillion Dollars by 2027," April 19, 2023, https://www.goldmansachs.com/intelligence/pages/the-creator-economy-could-approach-half-a-trillion-dollars-by-2027.html.

18. Jena McGregor, "Companies with Flexible Remote Work Policies Outperform on Revenue Growth: Report," *Forbes*, November 14, 2023, https://www.forbes.com/sites/jenamcgregor/2023/11/14/companies-with-flexible-remote-work-policies-outperform-on-revenue-growth-report/?sh=425263795ae4.

19. Kristen Senz, "How Companies Benefit When Employees Work Remotely," HBS Working Knowledge, July 29, 2019, https://hbswk.hbs.edu/item/how-companies-benefit-when-employees-work-remotely.

20. Philip Osafo-Kwaako et al., "Mobile Money in Emerging Markets: The Business Case for Financial Inclusion," McKinsey, March 12, 2018, https://www.mckinsey.com/industries/financial-services/our-insights/mobile-money-in-emerging-markets-the-business-case-for-financial-inclusion.

21. Ben Schecter, "The Future of Work Is Not Corporate—It's DAOs and Crypto Networks," a16zcrypto, December 17, 2021, https://a16zcrypto.com/posts/article/the-future-of-work-daos-crypto-networks/.

22. Glaveski, "How DAOs Could Change the Way We Work."

23. Ravin Jesuthasan and John W. Boudreau, *Work without Jobs: How to Reboot Your Organization's Work Operating System* (Cambridge, MA: MIT Press, 2022).

Chapter 2

1. "Roblox Developer Says She Made about $500,000 One Year Designing Games for the Platform," CNBC Make It, March 2024, https://www.linkedin.com/posts/cnbc-make-it_roblox-developer-says-she-made-about-500000-activity-7167951541578915840-SStR/.

2. Cecilia D'Anastasio, "Roblox Game Developers Earned a Record $741 Million Last Year," *Bloomberg*, February 21, 2024, https://www.bloomberg.com/news/articles/2024-02-21/how-much-do-roblox-game-makers-earn-741-million-in-2023.

3. Lavender Nguyen, "What Is Brand Transparency, and Why Is It Important in 2022?," Latana, August 29, 2022, https://resources.latana.com/post/brand-transparency/#:~:text=Transparency%20makes%20trust%20possible%2C%20and,loyal%20to%20a%20transparent%20brand.

4. Jim Harter, "U.S. Employee Engagement Needs a Rebound in 2023," Gallup, January 25, 2023, https://www.gallup.com/workplace/468233/employee-engagement-needs-rebound-2023.aspx#:~:text=This%20pattern%20continued%20into%202022,while%2018%25%20are%20actively%20disengaged.

5. "Tools & Samples," SHRM.org, March 2024, https://www.shrm.org/resourcesandtools/hr-topics/technology/pages/hr-new-technology-manage-contingent-workers.aspx#:~:text=Gartner%20research%20found%20that%20contingent,demand%20for%20niche%20digital%20skills.

6. Jay Bhatt, Colleen Bordeaux, and Jen Fisher, "The Workforce Well-Being Imperative," *Deloitte Insights*, March 13, 2023, https://www2.deloitte.com/us/en/insights/topics/talent/employee-wellbeing.html.

Chapter 3

1. Michelle Toh, "300 Million Jobs Could Be Affected by Latest Wave of AI, Says Goldman Sachs," CNN Business, March 29, 2023, https://www.cnn.com/2023/03/29/tech/chatgpt-ai-automation-jobs-impact-intl-hnk/index.html.

2. Goldman Sachs, "Generative AI Could Raise Global GDP by 7%," Intelligence, April 5, 2023, https://www.goldmansachs.com/intelligence/pages/generative-ai-could-raise-global-gdp-by-7-percent.html.

3. Toh, "300 Million Jobs Could Be Affected by Latest Wave of AI, Says Goldman Sachs."

4. Goldman Sachs, "Generative AI Could Raise Global GDP by 7%."

5. Toh, "300 Million Jobs Could Be Affected by Latest Wave of AI, Says Goldman Sachs."

6. Goldman Sachs, "Generative AI Could Raise Global GDP by 7%."

7. Toh, "300 Million Jobs Could Be Affected by Latest Wave of AI, Says Goldman Sachs."

8. Goldman Sachs, "Generative AI Could Raise Global GDP by 7%."

9. Toh, "300 Million Jobs Could Be Affected by Latest Wave of AI, Says Goldman Sachs."

10. Goldman Sachs, "Generative AI Could Raise Global GDP by 7%."

11. Toh, "300 Million Jobs Could Be Affected by Latest Wave of AI, Says Goldman Sachs."

12. Goldman Sachs, "Generative AI Could Raise Global GDP by 7%."

Chapter 4

1. Juicebox, "Juicebox Enables the Creators of Tomorrow to Launch, Fund and Manage the Boldest Projects on the Internet," n.d., https://juicebox.money/about.

2. Sina Habibian, *Jango & Nnnnicholas: Juicebox, Programming Internet-Native Organizations,* podcast, September 28, 2022, https://www.intothebytecode.com/jango-nnnnicholas-juicebox/.

3. "Web 3.0 Market Worth $5.5 Billion by 2030—Exclusive Report by MarketsandMarkets," PR Newswire, October 11, 2023, https://www.prnewswire.com/news-releases/web-3-0-market-worth-5-5-billion-by-2030—exclusive-report-by-marketsandmarkets-301953324.html.

4. Rajesh Kandaswamy, David Furlonger, and Andrew Stevens, "Digital Disruption Profile: Blockchain's Radical Promise Spans Business and Society," Gartner Research, n.d., https://www.gartner.com/en/doc/3855708-digital-disruption-profile-blockchains-radical-promise-spans-business-and-society.

5. Harvard Business School, "Harvard Business School Professor Wickham

Skinner Dies at 94," press release, February 6, 2019, https://www.hbs.edu/news/releases/Pages/c-wickham-skinner-obituary.aspx.

6. Deloitte, Workplace Burnout Survey, n.d., https://www2.deloitte.com/us/en/pages/about-deloitte/articles/burnout-survey.html; Ben Wigert and Sangeeta Agrawal, "Employee Burnout, Part 1: The 5 Main Causes," Gallup, July 12, 2018, https://www.gallup.com/workplace/237059/employee-burnout-part-main-causes.aspx.

7. Solidity, company website, n.d., https://soliditylang.org/.

8. OECD, "Managing Decentralisation: A New Role for Labour Market Policy, Local Economic and Employment Development (LEED)," 2003, OECD Publishing, https://www.oecd-ilibrary.org/employment/managing-decentralisation_9789264104716-en.

9. Medrec:M, "A Modern Way to Track Your Health," company website, n.d., https://medrec-m.com/.

10. Deloitte, "Tech Trends 2024," Deloitte Insights, n.d., https://www2.deloitte.com/us/en/insights/focus/tech-trends.html.

11. Isabelle Lee, "ConstitutionDAO Disbands after Losing Its Bid to Buy a Copy of the Constitution," *Business Insider*, November 24, 2021, https://markets.businessinsider.com/news/currencies/constitution-dao-disbands-refund-how-us-constitution-discord-website-twitter-2021-11.

12. McKinsey & Company, "2021 Top Picks: From Recovery to Growth," n.d., https://www.mckinsey.com/~/media/mckinsey/business%20functions/marketing%20and%20sales/our%20insights/2021%20top%20picks%20from%20recovery%20to%20growth/2021-top-picks-from-recovery-to-growth.pdf.

13. Wendy Henry, et al., "Using Blockchain to Drive Supply Chain Transparency," Deloitte, June 2023, https://www2.deloitte.com/us/en/pages/operations/articles/blockchain-supply-chain-innovation.html.

14. George Westerman et al., "The Digital Advantage: How Digital Leaders Outperform Their Peers in Every Industry," Capgemini Consulting and *MIT Sloan Management Review*, July 2017, https://www.capgemini.com/wp-content/uploads/2017/07/The_Digital_Advantage__How_Digital_Leaders_Outperform_their_Peers_in_Every_Industry.pdf.

15. MIT SMR Strategy Forum, "Is Blockchain a Disruptive or a Sustaining Innovation? What Experts Say," *MIT Sloan Management Review*, March 30, 2022, https://sloanreview.mit.edu/strategy-forum/is-blockchain-a-disruptive-or-a-sustaining-innovation-what-experts-say/.

Chapter 5

1. Interview with authors.

2. Interview with authors.

3. Matthew Ball, *The Metaverse: And How It Will Revolutionize Everything* (New York: Liveright, 2022).

4. Meta, "Introducing the New Ray-Ban/Meta Smart Glasses," press release, September 27, 2023, https://about.fb.com/news/2023/09/new-ray-ban-meta-smart-glasses/.

5. Dan Carney, "NIVIDIA Omniverse Brings BMW a 30 Percent Boost in Production Planning Efficiency," *Design News*, May 17, 2021, https://www.designnews.com/design-software/nvidia-omniverse-brings-bmw-30-percent-boost-production-planning-efficiency#:~:text=%E2%80%9CThe%20 capability%20to%20operate%20in,percent%20more%20efficient%2C%20he%20sai.

6. Richard Kerris, "Ericsson Builds Digital Twins for 5G Networks in NVIDIA Omniverse," *NIVIDIA* (blog), November 9, 2021, https://blogs.nvidia.com/blog/2021/11/09/ericsson-digital-twins-omniverse/.

7. McKinsey & Company, "What Is Personalization?," May 30, 2023, https://www.mckinsey.com/featured-insights/mckinsey-explainers/what-is-personalization.

8. Melody Brue, "The Rise of Augmented Reality in the Modern Workplace," *Forbes*, June 19, 2023, https://www.forbes.com/sites/moorinsights/2023/06/19/the-rise-of-augmented-reality-in-the-modern-workplace/?sh=37a20c342ad4.

9. Philipp Sostmann, LinkedIn, 2024, https://www.linkedin.com/in/pasostmann/.

10. PwC, "Virtual and Augmented Reality Could Deliver a $1.5 Trillion Boost to the Global Economy by 2030—PwC," press release, January 29, 2020, https://www.pwc.com/th/en/press-room/press-release/2020/press-release-29-01-20-en.html.

11. Jane Incao, "How VR Is Transforming the Way We Train Associates," *Walmart Today*, September 20, 2018, https://corporate.walmart.com/news/2018/09/20/how-vr-is-transforming-the-way-we-train-associates#:~:text=We've%20also%20seen%20that,saw%20the%20same%20retention%20boosts.%E2%80%9D.

12. Brie Weiler Reynolds, "5 Reasons Remote Work Is Most Definitely on the Rise," Flexjobs, n.d., https://www.flexjobs.com/blog/post/reasons-remote-work-is-not-in-decline/#:~:text=The%20long%2Dterm%20growth%20of,in%20the%20last%20two%20years.

13. Interview with Monzon.

Chapter 6

1. Bessie Liu, "Kevin Owocki on Building a 'Pro-Topian' Future for Blockchains," *Blockworks*, October 12, 2023, https://blockworks.co/news/kevin-owocki-permissionless-gitcoin.

2. "Facebook Buys Instagram for $1 Billion, Turns Budding Rival into Its Standalone Photo App," TechCrunch, April 9, 2012, https://techcrunch.com/2012/04/09/facebook-to-acquire-instagram-for-1-billion.

3. Versr, "Service DAOs," Dcentral, https://daocentral.com/explore/service.

4. Crytopedia Staff, "What Was the DAO?," Crytopedia, October 5, 2023, https://www.gemini.com/cryptopedia/the-dao-hack-makerdao.

5. David Siegel, "Understanding the DAO Attack," CoinDesk, June 25, 2016, https://www.coindesk.com/learn/understanding-the-dao-attack/.

6. Phil Daian, "Analysis of the DAO Exploit," *Hacking, Distributed* (blog), June 18, 2016, https://hackingdistributed.com/2016/06/18/analysis-of-the-dao-exploit/.

7. Blockchain Academy, *The Global Blockchain Employment Report*, 2021, https://theblockchaintest.com/uploads/resources/the%20Blockchain%20Academy%20-%20the%20Global%20Blockchain%20Employment%20Report%20-%202022%20March.pdf.

8. Pagan Kennedy, "William Gibson's Future Is Now," *New York Times*, January 13, 2012, https://www.nytimes.com/2012/01/15/books/review/distrust-that-particular-flavor-by-william-gibson-book-review.html.

Chapter 7

1. Evan Andrews, "The History of the Handshake," History, September 12, 2023, https://www.history.com/news/what-is-the-origin-of-the-handshake.

2. Andrew O'Neill et al., "What Can You Trust in a Trustless System?" S&P Global, October 11, 2023, https://www.spglobal.com/en/research-insights/featured/special-editorial/what-can-you-trust-in-a-trustless-system.

3. Talentpair, "Talentpair Wins 'Overall Recruiting Solution of the Year' for 2023 RemoteTech Breakthrough Awards," press release, June 22, 2023, https://talentpair.com/featured/talentpair-wins-overall-recruiting-solution-of-the-year-for-2023-remotetech-breakthrough-awards/.

4. Victoria Li, "Smart Contracts for HR: Automating Hiring Processes in DeFi Companies," Future Jobs by Cointelegraph, January 4 (no year), https://jobs.cointelegraph.com/blog/smart-contracts-for-hr-hiring-processes-in-defi.

5. Office of Public Affairs, "Union Election Petitions Increase 57% in First Half of Fiscal Year 2022," National Labor Relations Board, press release, April 6, 2022, https://www.nlrb.gov/news-outreach/news-story/union-election-petitions-increase-57-in-first-half-of-fiscal-year-2022.

6. Megan Brenan, "Approval of Labor Unions at Highest Point Since 1965," Gallup, September 2, 2021, https://news.gallup.com/poll/354455/approval-labor-unions-highest-point-1965.aspx.

7. Alina Selyukh, "Starbucks Union Push Spreads to 54 Stores in 19 States," NPR, January 31, 2022, https://www.npr.org/2022/01/31/1076978207/starbucks-union-push-spreads-to-54-stores-in-19-states.

8. Sara Ashley O'Brien, "Amazon Warehouse Workers in New York Made History Voting for a Union. Here's What Could Happen Next," CNN Business, April 4, 2022, https://www.cnn.com/2022/04/04/tech/amazon-labor-union-staten-island-whats-next/index.html.

9. FBI, "Tornado Cash Co-Founders Accused of Helping Cybercriminals Launder Stolen Crypto," FBI News, September 7, 2023, https://www.fbi.gov/news/stories/tornado-cash-co-founders-accused-of-helping-cybercriminals-launder-stolen-crypto.

10. PwC, "Five Ways HR Leaders Can Make the Most of Their Technology Investments," PC HR Tech Survey 2022, February 3, 2022, https://www.pwc.com/us/en/tech-effect/cloud/hr-tech-survey.html.

Chapter 8

1. Marcos Rezende, personal website, https://www.marcosrezende.com/.

2. Interview with author, videoconference call, February 12, 2024.

3. Interview with author, videoconference call.

4. Josh Howarth, "57+ Freelance Statistics, Trends and Insights (2024)," Exploding Topics, February 19, 2024, https://explodingtopics.com/blog/freelance-stats.

5. Upwork, "Upwork Unveils Most In-Demand Work Skills in 2024," press release, March 19, 2024, https://investors.upwork.com/news-releases/news-release-details/upwork-unveils-most-demand-work-skills-2024.

6. Vanmala Subramaniam, "Apps like Uber and DoorDash Use AI to Determine Pay. Works Say This Makes It Impossible to Predict Wages," *Globe and Mail*, July 8, 2023, https://www.theglobeandmail.com/business/article-pay-ai-algorithm-gig-workers-uber/.

7. Ben Zipperer et al., "National Survey of Gig Workers a Picture of Poor Working Conditions, Low Pay," Economic Policy Institute, June 1, 2022, https://www.epi.org/publication/gig-worker-survey/.

8. Dominik Metelski and Janusz Sobieraj, "Decentralized Finance (DeFi) Projects: A Study of Key Performance Indicators in Terms of DeFi Protocols' Valuations," *International Journal of Financial Studies* 10, no. 4 (2022): 108, https://www.mdpi.com/2227-7072/10/4/108.

9. Deloitte, "Using Blockchain to Drive Supply Chain Transparency," *Deloitte Insights*, n.d., https://www2.deloitte.com/us/en/pages/operations/articles/blockchain-supply-chain-innovation.html.

10. Chainlink, "What Is a Blockchain Oracle?," Education, January 12, 2024, https://chain.link/education/blockchain-oracles.

11. Felipe Erazo, "100M People Worldwide Now Use Crypto-Based Assets, Says Cambridge Study," Cointelegraph, September 24, 2020, https://cointelegraph.com/news/100m-people-worldwide-now-use-crypto-based-assets-says-cambridge-study.

12. McKinsey, "Rekindling US Productivity for a New Era," McKinsey Global Institute, February 16, 2023, https://www.mckinsey.com/mgi/our-research/rekindling-us-productivity-for-a-new-era#introduction.

13. LinkedIn, "Workplace Learning Report," LinkedIn Learning, 2021, https://learning.linkedin.com/content/dam/me/business/en-us/amp/learning-solutions/images/wlr21/pdf/LinkedIn-Learning_Workplace-Learning-Report-2021-EN-1.pdf.

14. Boston Consulting Group, "Managing Work and the Workforce in Health Care's New Reality," January 2022, https://web-assets.bcg.com/f2/4c/eff0ccbe453eb5b224159d118ecb/bcg-managing-work-and-the-workforce-in-health-cares-new-reality-jan-2022-r.pdf.

15. World Bank, "Record High Remittances Sent Globally in 2018," press release, April 8, 2019, https://www.worldbank.org/en/news/press-release/2019/04/08/record-high-remittances-sent-globally-in-2018.

16. Chainlink, "What Is Layer 2?," Education, May 24, 2023, https://chain.link/education-hub/what-is-layer-2#:~:text=A%20layer%202%20refers%20to,such%20as%20higher%20transaction%20throughputs.

Chapter 9

1. Jay Buolamwini and Timrit Gebru," Gender Shades: Intersectional Accuracy Disparities in Commercial Gender Classification," *In Proceedings of the 1st Conference on Fairness, Accountability and Transparency* PMLR, 2018, 77–91, https://proceedings.mlr.press/v81/buolamwini18a.html.

2. Emily Bender et al.," On the Dangers of Stochastic Parrots: Can Language Models Be Too Big? "In *Proceedings of the 2021 ACM Conference on Fairness, Accountability, and Transparency*, 2021, 610–623, https://doi.org/10.1145/3442188.3445922.

3. Violet Turri, "What Is Explainable AI?," *SEI Blog*, Software Engineering Institute, Carnegie Mellon University, January 17, 2022, https://insights.sei.cmu.edu/blog/what-is-explainable-ai/.

4. Scott Reyburn, "JPG File Sells for $69 Million, as 'NFT Mania' Gathers Pace," *New York Times*, March 11, 2021, https://www.nytimes.com/2021/03/11/arts/design/nft-auction-christies-beeple.html.

5. Ahmed Elgammal, "What the Art World Is Failing to Grasp about Christie's AI Portrait Coup," Artsy, October 29, 2018, https://www.artsy.net/article/artsy-editorial-art-failing-grasp-christies-ai-portrait-coup.

6. Isaiah Poritz, "Open AI Legal Troubles Mount with Suit over AI Training on Novels," Bloomberg Law, June 29, 2023, https://news.bloomberglaw.com/ip-law/openai-facing-another-copyright-suit-over-ai-training-on-novels.

7. "'AI Is All Around Us'—Neil deGrasse Tyson Says We Shouldn't Worry about Artificial Intelligence," *Late Show with Stephen Colbert* (via YouTube), https://youtu.be/5Qon72VKH30?feature=shared.

8. *Thaler (Appellant) v. Comptroller-General of Patents, Designs and Trademarks* (Respondent), case 2021/0201, British Supreme Court, March 2, 2023, https://www.supremecourt.uk/cases/uksc-2021-0201.html. https://www.federalregister.gov/documents/2024/02/13/2024-02623/inventorship-guidance-for-ai-assisted-inventions#:~:text=Further%2C%20the%20USPTO%20recognizes%20that,constitute%20inventorship%20under%20our%20laws.

9. Lorepunk, "What We Know about the Contract Vulnerability Worrying Web3," NFTNow, December 5, 2023, https://nftnow.com/news/what-we-know-about-the-contract-vulnerability-worrying-web3/.

Chapter 10

1. Gallup, "Hybrid Work," n.d., https://www.gallup.com/401384/indicator-hybrid-work.aspx.

2. Te-Ping Chen, "Remote Workers Are Losing Out on Promotions, New Data Shows," *Wall Street Journal*, January 24, 2024, https://www.wsj.com/lifestyle/careers/remote-workers-are-losing-out-on-promotions-8219ec63.

3. Jordan Turner, "Employees Seek Personal Value and Purpose at Work. Be Prepared to Deliver," Gartner, March 29, 2023, https://www.gartner.com/en/articles/employees-seek-personal-value-and-purpose-at-work-be-prepared-to-deliver.

4. Marcus Buckingham, *Love + Work* (Boston: Harvard Business Review Press, 2022).

5. Buffer, "State of Remote Work 2023," Buffer, https://buffer.com/state-of-remote-work/2023.

6. PwC, "Virtual and Augmented Reality Could Deliver a £1.4 Trillion Boost to the Global Economy by 2030—PwC," press release, n.d., https://www.pwc.com/id/en/media-centre/press-release/2020/english/virtual-and-augmented-reality-could-deliver-a-p1-4trillion-boost.html.

7. Stanford, "Stanford Study Finds Walking Improves Creativity," *Stanford Report*, April 24, 2014, https://news.stanford.edu/2014/04/24/walking-vs-sitting-042414/.

8. r/OculusQuest, "Can Meta Quest 3 Be Used by the Disabled," Reddit, December 2023, https://www.reddit.com/r/OculusQuest/comments/1829wgd9/can_meta_quest_3_be_used_by_the_disabled/.

9. Dash Bibhudatta, "Generative AI Will Transform Virtual Meetings," hbr.org, November 29, 2023, https://hbr.org/2023/11/generative-ai-will-transform-virtual-meetings.

10. MIT SMR Connections, "Creating a Connected Culture: Strategies for Enhancing Inclusion and Engagement," *MIT Sloan Management Review*, May 24, 2023, https://sloanreview.mit.edu/mitsmr-connections/creating-a-connected-culture-strategies-for-enhancing-inclusion-and-engagement/.

11. Inside Track staff, "A Foundation for Modern Collaboration: Microsoft 365 Bolsters Teamwork," Microsoft, February 28, 2024, https://www.microsoft.com/insidetrack/blog/a-foundation-for-modern-collaboration-microsoft-365-bolsters-teamwork/.

12. Travis Howell, "Coworking Spaces: An Overview and Research Agenda," *Research Policy* 51, no. 2 (March 2022), https://www.sciencedirect.com/science/article/pii/S0048733321002390.

13. Ben Schecter, "The Future of Work Is Not Corporate—It's DAOs and Networks," a16zcrypto, December 17, 2021, https://a16zcrypto.com/content/article/the-future-of-work-daos-crypto-networks/.

14. MBO Partners, "Digital Nomads," n.d., https://www.mbopartners.com/state-of-independence/digital-nomads/.

15. PwC, "What Does Virtual Reality and the Metaverse Mean for Training?," n.d., https://www.pwc.com/us/en/tech-effect/emerging-tech/virtual-reality-study.html#:~:text=In%20fact%2C%20learners%20trained%20with,improvement%20over%20e%2Dlearn%20training.

16. Upwork, "Economist Report: Remote Workers on the Move," press release, n.d., https://www.upwork.com/press/releases/economist-report-remote-workers-on-the-move#:~:text=Remote%20work%20will%20increase%20migration,times%20what%20they%20normally%20are.

17. Mara Pérez, "Patagonian's Company Culture: Their Way to Employee Retention," *Nailted* (blog), March 25, 2022, https://nailted.com/blog/patagonians-company-culture-their-own-way-to-employee-retention/.

18. Bryan Robinson, "'Chronoworking': Another Career Trend to Achieve Work-Life Balance in 2024," *Forbes*, April 9, 2024, https://www.forbes.com/sites/bryanrobinson/2024/04/09/chronoworking-another-career-trend-to-achieve-work-life-balance-in-2024/?sh=13321855642e.

19. Bill Chappell, "4-Day Workweek Booster Workers' Productivity by 40%, Microsoft Japan Says," *All Things Considered*, NPR, November 4, 2019, https://www

.npr.org/2019/11/04/776163853/microsoft-japan-says-4-day-workweek-boosted-workers-productivity-by-40#:~:text=store%20in%20Tokyo.-,Microsoft's%20division%20in%20Japan%20says%20it%20saw%20productivity%20grow%20by,a%20week%20rather%20than%20five.&text=Workers%20at%20Microsoft%20Japan%20enjoyed,normal%2C%20five%2Dday%20paycheck.

20. Business Group on Health, "77% of Employers Report Increase in Workforce Mental Health Needs, Says Business Group on Health's 2024 Health Care Strategy Survey," press release, August 22, 2023, https://www.businessgrouphealth.org/en/newsroom/news%20and%20press%20releases/press%20releases/2024%20lehcss.

21. Gitlab, "Family and Friends Day," GitLab Handbook, n.d., https://handbook.gitlab.com/handbook/company/family-and-friends-day/.

22. Maria Roche and Andy Wu, "What's the Optimal Workplace for Your Organization," hbr.org, February 9, 2022, https://hbr.org/2022/02/whats-the-optimal-workplace-for-your-organization.

23. Kirsten Weir, "Give Me a Break," *Monitor on Psychology* 50, no. 1: (January 2019), https://www.apa.org/monitor/2019/01/break.

24. Gallup, "The Benefits of Employee Engagement," Gallup Workplace, June 20, 2013, https://www.gallup.com/workplace/236927/employee-engagement-drives-growth.aspx.

Chapter 11

1. Melanie Hanson, "Average Cost of College & Tuition," Education Data Initiative, November 18, 2023, https://educationdata.org/average-cost-of-college#:~:text=The%20average%20cost%20of%20attendance,or%20%24223%2C360%20over%204%20years.

2. Accenture, Grads for Life, and Harvard Business School, "Dismissed by Degrees," December 13, 2017, https://www.hbs.edu/managing-the-future-of-work/Documents/dismissed-by-degrees.pdf.

3. Susan Milligan and Lauren Camera, "Ditch the Degree? Many Employers Are Fine with That," *US News*, February 2, 2023, https://www.usnews.com/news/the-report/articles/2023-02-03/ditch-the-degree-many-employers-are-just-fine-with-that#:~:text=Research%20shows%20that%20a%20degree,are%20reevaluating%20the%20old%20standards.

4. Kristen Sze, "Bay Area High School Grad Rejected by 16 Colleges Hired by Google," ABC News, October 11, 2023, https://abc7news.com/stanley-zhong-college-rejected-teen-full-time-job-google-admissions/13890332/.

5. Lex Borghans et al., "What Grades and Achievement Tests Measure," IZA Discussion Paper no. 10356, November 2016, https://docs.iza.org/dp10356.pdf.

6. Kylie Ora Lobell, "SHRM: Why Fewer Employers Are Requiring College Degrees," Burning Glass Institute, September 11, 2023, https://www.burningglassinstitute.org/news/shrm-why-fewer-employers-are-requiring-college-degrees.

7. Rachel M. Cohen, "Stop Requiring College Degrees for Jobs That Don't Need Them," Vox, March 19, 2023, https://www.vox.com/policy/23628627/degree-inflation-college-bacheors-stars-labor-worker-paper-ceiling.

8. Rachel M. Cohen, Vox, March 19, 2023.

9. "The Tech Agenda," PwC, accessed May 2024, https://www.pwc.com/gx/en/industries/technology/publications/blockchain-report-transform-business-economy.html; Sara Castellanos, "PwC Tests Blockchain for Validating Job Candidates' Credentials," Wall Street Journal, April 3, 2019, https://www.wsj.com/articles/pwc-tests-blockchain-for-validating-job-candidates-credentials-11554324777.

10. Pratik Mistry, "How AI Is Responsible for the Transformation of the Education Industry," eLearning Industry, January 27, 2023, https://elearningindustry.com/how-ai-is-responsible-for-the-transformation-of-the-education-industry.

11. Pattie Maes and Joanne Leong, "Evaluating the Effectiveness of AI-Generated Personalized Learning Content for Improving Engagement and Learning Outcomes," Integrated Learning Initiative, MIT Media Lab, June 30, 2024, https://mitili.mit.edu/research/evaluating-effectiveness-ai-generated-personalized-learning-content-improving-engagement.

12. Gartner Peer Community, "What Steps Would You Take as a Leader to Create Continuous Learning and Growth Culture within Your Team?," Gartner, May 2023, https://www.gartner.com/peer-community/post/steps-take-leader-to-create-continuous-learning-growth-culture-within-team.

13. World Economic Forum, "Towards a Reskilling Revolution," World Economic Forum and Boston Consulting Group, January 2018, https://www3.weforum.org/docs/WEF_FOW_Reskilling_Revolution.pdf.

14. Career Karma, "State of the Bootcamp Market Report 2020," https://careerkarma.com/blog/wp-content/uploads/2020/02/State-Of-The-Bootcamp-Market-Report-2020.pdf.

15. "Compte Personnel de Formation," Alliance Française de Paris, https://www.alliancefr.org/en/cpf.

16. Deloitte, "Soft Skills for Business Success," Deloitte Perspective, n.d., https://www.deloitte.com/au/en/services/economics/perspectives/soft-skills-business-success.html.

Chapter 12

1. TriNet, "People Operations: A Guide to the New HR Term You Should Know," TriNet Insights, December 4, 2023, https://www.trinet.com/insights/people-operations.

2. Deloitte, "Unleashing Value from Digital Transformation: Paths and Pitfalls," Deloitte Research, February 14, 2023, https://www.deloitte.com/global/en/our-thinking/insights/topics/digital-transformation/digital-transformation-value-roi.html.

3. Gerald C. Kane et al., "Aligning the Organization for Its Digital Future," *MIT Sloan Management Review* and Deloitte University Press, Summer 2016, https://www2.deloitte.com/content/dam/insights/us/articles/mit-smr-deloitte-digital-transformation-strategy/2016_MIT_DeloitteAligningDigitalFuture.pdf.

4. Rebecca King, "Web3: The Hype and How It Can Transform the Internet," World Economic Forum, February 1, 2022, https://www.weforum.org/agenda/2022/02/web3-transform-the-internet/.

5. "State of the Global Workplace," Gallup, 2024, https://www.gallup.com/workplace/349484/state-of-the-global-workplace.aspx.

6. "What Is Microsoft Mesh? Creating Immersive Workspaces," *UC Today* (blog), November 20, 2023, https://www.uctoday.com/collaboration/what-is-microsoft-mesh/.

7. "What Does Virtual Reality and the Metaverse Mean for Training?," PwC, September 15, 2022, https://www.pwc.com/us/en/tech-effect/emerging-tech/virtual-reality-study.html.

Conclusion

11. Accenture, "Care to Do Better," Accenture research report, September 23, 2020, https://www.accenture.com/us-en/insights/future-workforce/employee-potential-talent-management-strategy.

INDEX

ACKNOWLEDGMENTS

Life sometimes presents pivotal moments that demand courage—moments when you find yourself at the edge of an intellectual cliff, faced with the decision to leap into the unknown. The creation of this book represents one such daring plunge.

Before we embarked on this literary journey, our paths had not crossed. Our sole interaction was when Josh applied to speak at a web3 conference co-organized by Deborah. Yet, in that leap of faith when Deborah picked up the phone to reach out to Josh despite the dozens of other speaking applications she'd received, we discovered a profound connection: a shared, deep-rooted passion for democratizing opportunity and revolutionizing work as we know it.

In the era of AI and web3 technologies, we're witnessing the dawn of a revolutionary work paradigm. This new landscape offers unprecedented potential for worker empowerment, engagement, and fulfillment. Our mission is to guide organizations through this transformative journey, helping them harness these technologies to create more meaningful, equitable, and rewarding work environments for all.

Both of us had a vision into the future. Deborah was inspired by her Gen Z children, who, in turn, were inspired by the parents of their Silicon Valley friends—pioneers at the forefront of the AI and web3 revolution. And Josh, who spent long days and late nights working with young, hungry startups and students at the Harvard Innovation Labs, was inspired by their new and visionary methods for organizing work today and into the future. The future belongs to this emerging generation and its ambitious goals and dreams.

Combining our complementary expertise, we embarked on the collaborative journey of writing this book. The process, like the subject matter itself, exemplifies the power of shared knowledge and teamwork. From the moment we met in person for the very first time at a writing retreat in Napa Valley, California, to the penning of the last word, we were amazed at the invaluable insights and unwavering dedication we received on this project. Writing a book is a collaborative journey, and we are immensely grateful to those who have supported us throughout the development of *Employment is Dead: How Disruptive Technologies Are Revolutionizing the Way We Work.*

A special thanks to Kevin Evers, senior editor at Harvard Business Review Press, who took a leap of faith on our proposal and acquired our book, and was willing to bend and adapt as the speed of rapid change in AI and web3 took our breath away. Then there's our editor at the Press, Courtney Schinke Cashman, whose keen eye and thoughtful guidance have shaped this manuscript from its inception. Courtney, your commitment to clarity and precision has been instrumental in bringing our ideas to life.

We are also grateful to several other members of the remarkable HBR Press team, including Felicia Sinusas, Julie Devoll, Sally Ashworth, Victoria Desmond, Jon Shipley, Jordan Concannon, Alexandra Kephart, Cheyenne Paterson, and Lindsey Dietrich.

We also extend our gratitude to our colleagues and partners at the Work3 Institute, whose thought-provoking discussions and diverse perspectives have continually inspired and challenged us to think beyond the conventional boundaries of work. We are deeply appreciative of our research assistants, Andrea Li and Chie Davis, for their meticulous work in gathering data and providing critical analysis. Your tireless efforts have helped ensure the accuracy and depth of our content.

Our heartfelt thanks go to the leaders and innovators who graciously shared their stories and experiences, offering a glimpse into the transformative potential of web3, blockchain, and AI: Val Bercovici, Edo Segal, Jim Kaskade, Eron Sunando, Neha Singh, Richard Kerris, Tracie Sponenberg, Mark Christensen, Tyson Snelson, Philipp Sostmann, Patrik Drean, Jason Palmer, Donna Scarola, and Julie DeMartini. Your contributions have added real-world context and richness to our narrative.

To the six children we have between us, Drake, Dominick, Dayne, Tristan, Liam, and Beckham, we appreciate your patience while we were writing, deep in thought, or overly excited about this work. We also want to thank our spouses, Dino and Lauren, who served as sounding boards and took on extra responsibilities while we were in our writing caves. We couldn't have done this without your love and support. To our many friends, thank you for your encouragement, advice, and belief in our vision. Your unwavering support has been a cornerstone of our journey.

To our agent, Claudia Cross, at Folio Literary Management, we extend our heartfelt gratitude for your unending support, expert guidance, and steadfast belief in us and in this project. Your experience has been invaluable in bringing this book to fruition.

Finally, to our readers, thank you for joining us on this exploration of the future of work. We hope this book sparks conversation, inspires innovation, and encourages you to embrace the possibilities that lie ahead.

With gratitude,
Deborah Perry Piscione and Josh Drean

ABOUT THE AUTHORS

DEBORAH PERRY PISCIONE moved to Silicon Valley in 2006. Shortly after relocating, while standing in line at a local Starbucks, the woman ahead of her noticed she was new to the area and asked, "How can I help you?" This conversation led to a meeting with a leading venture capitalist at a top-tier venture capital firm, which later resulted in another introduction and a $5 million investment in Perry Piscione's first startup. Her career trajectory had been dramatically altered.

Perry Piscione was deeply impressed by Silicon Valley's culture, which stood in stark contrast to her previous work experiences. Having worked in foreign policy in the halls of Congress and the White House and served as a television commentator on CNN, MSNBC, and Fox News in Washington, DC, she was accustomed to an environment where division was the norm. Silicon Valley's collaborative spirit opened her eyes to a new way of thinking and doing business.

Captivated by Silicon Valley's culture Perry Piscione wrote the *New York Times* bestseller and highly acclaimed book, *Secrets of Silicon Valley*, which was acquired in thirty-nine countries. A leading global consulting firm adopted the book to build out the firm's innovation boot-camp practice. From this experience, Piscione architected Improvisational Innovation, which helps companies innovate using a bottom-up approach in which anyone in any corner of the company could introduce new ideas in a trusted and safe environment. This approach has now been adopted by many *Fortune* 500 companies.

One of Perry Piscione's proudest accomplishments is founding the Alley to the Valley (A2V) community. This innovative network quickly earned a reputation as a "female golf course," where accomplished women could forge powerful connections and close deals, through the

A2V ask and offer methodology. Through this initiative, Perry Piscione witnessed firsthand the transformative power of networking spaces that fast-track dealmaking among accomplished women, fostering high-level collaboration.

Perry Piscione's LinkedIn Learning courses, "Risk Taking for Leaders" and "Executing on Innovation," have been translated into a dozen languages, reaching a global audience eager for her insights. As a sought-after global speaker, she shares her knowledge of Work3 transformation, the Silicon Valley ecosystem, innovation, and risk-taking leadership. Recognizing her significant contributions, Stanford University's Graduate School of Business conducted a case study titled "Deborah Perry Piscione: Finding Opportunity in Silicon Valley," highlighting her status as a thought leader in entrepreneurship and innovation.

Together, as renowned experts in AI, web3 technologies, and the future of work, Perry Piscione and Josh Drean cofounded the Work3 Institute. This pioneering research and advisory firm equips organizations with innovative frameworks to navigate the seismic shift from traditional work structures to the new paradigms demanded by the AI and web3 era. Drawing on their extensive experience and visionary insights, they guide companies through the complex process of reimagining and restructuring their operations for a rapidly evolving technological landscape.

When she is not working to democratize opportunity, Perry Piscione sits on advisory and corporate boards, is an advocate for foster youth, and enjoys the California lifestyle with her three children, Drake, Dominick, and Dayne and their mini goldendoodle, Duchess.

JOSH DREAN is a renowned expert in talent management, people analytics, and workforce experience, with a deep passion for integrating these disciplines with emerging technologies. As cofounder of the Work3 Institute, Drean is at the forefront of innovative and forward-thinking approaches to enhancing workforce strategies. His role as a workforce advisor at the Harvard Innovation Labs further highlights his dedication to transforming workplace dynamics and ensuring that companies are well equipped to navigate the future of work.

Drean's journey began with his ten years as a youth motivational speaker, inspiring young audiences to overcome life's challenges with resilience and optimism. This work was the foundation for his later role as a web3 and workforce adviser to emerging companies and now as cofounder of the Work3 Institute.

With over a decade of experience, Drean has helped organizations across industries understand and adapt to the unique needs of millennials and Gen Z employees. His expertise lies in bridging generational divides, fostering inclusive and innovative work environments, and empowering employees to drive organizational success.

Drean's unique ability to connect with people through humor and storytelling has made him a highly sought-after speaker and consultant, working with companies ranging from ambitious startups to established *Fortune* 500 organizations. He actively engages with the emerging workforce, addressing their needs, preferences, and concerns on his social media channels, which garner millions of views annually.

Drean lives in Minneapolis, Minnesota, with his wife and their three children. He continues to push the boundaries of what it means to lead and succeed in the modern world of work.